Communicating in School Science

Communicating in School Science:
Groups, Tasks and Problem Solving 5–16

Di Bentley and Mike Watts

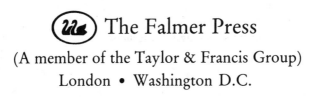
The Falmer Press
(A member of the Taylor & Francis Group)
London • Washington D.C.

UK	The Falmer Press, 4 John Street, London, WC1N 2ET
USA	The Falmer Press, Taylor & Francis Inc., 1900 Frost Road, Suite 101, Bristol, PA 19007

© D. Bentley and M. Watts 1992

All rights reserved. No part of this publication may be reproduced, stored in a retrieval system, or transmitted in any form or by any means, electronic, mechanical, photocopying, recording or otherwise, without permission in writing from the Publisher.

First published in 1992

British Library Cataloguing in Publication Data

Bentley, Di
  Communicating in school science.
  I. Title II. Watts, Mike
  507

  ISBN 1 85000 642 3
  ISBN 1 85000 643 1 (pbk)

Library of Congress Cataloging-in-Publication Data available on request

Jacket design by Caroline Archer

Typeset in 10/12 pt Garamond
Graphicraft Typesetters Ltd., Hong Kong

Printed by Burgess Science Press,
Rankine Road, Basingstoke, Hampshire

Contents

About the Authors	vi
Preface	vii
1 Communicating for Understanding	1
2 Creating a Climate for Communication	27
3 Communication and Group Work	56
4 National Curriculum Tasks in the Classroom	92
5 Communication in Open-ended Problem-solving	112
6 Communicating Achievement: Assessment in Science	138
References	183
Author and Subject Index	191

About the Authors

Di Bentley is Senior Adviser for Monitoring and Evaluation with Buckinghamshire County Council. After a decade of teaching in inner-city comprehensive schools in Manchester she became Research and Development Officer for Health Education as part of the Secondary Science Curriculum Review. Then, prior to its demise, she was an IBIS Inspector for the Inner London Education Authority. She has written widely in health and science education, her most recent books tackle teaching for active learning, and methods of assessing coursework in science.

Mike Watts is Reader in Education at Roehampton Institute, London. From teaching in inner-city comprehensive schools in London and Jamaica, he undertook research at the University of Surrey concerning children's understanding of concepts in physics. With the Secondary Science Curriculum Review he developed interests in many aspects of teaching and learning in science — his current role embraces a broad range of research issues in teacher education. He has published widely and is the author, with colleagues, of books and materials on various aspects of education, most recently books on open-ended problem-solving, and cross-curricular work in science.

Preface

Most of us have been assured in times gone by that school days are the happiest days of our lives — a notion that generally throws a distinct pall over any future prospects. But then, in those days we were not blessed with the National Curriculum. The late-eighties will be an educational benchmark in the UK for many generations to come — the 1988 Education Act is clearly twice as far-reaching as the 1944 Act so that we can only wait in awe for the year 2032. The 1988 Act has re-written many aspects of teachers' professional lives, explicitly and implicitly, and many of those changes are yet to become apparent.

In science education we have the pleasing prospect of seeing some cherished ideals and ideas brought within the ambit of all children from 5 to 16. Children into the next century will be provided with opportunities in science which were once the privilege of the very few. There are not all of the ideals all would wish for but — arguably — there are now included in school science many of the ideals most of us would want. For ourselves, one of these is a due emphasis on — and recognition of — the role of communicating in science. It has long been one of the central aims of science education that youngsters should attain a basic understanding of the world we live in, and be able to communicate that understanding to others. The role of communication in science has increasingly become an integral part of school science syllabuses. Perhaps teachers have always been aware of the need to feed children the opportunity to write, talk and act science. If so, the current emphasis will be due reward for their efforts. If not, and we suspect not, then the moment has arrived when we can begin to draw attention to a need.

Our intention in the book is to examine communication in classroom processes as it relates to both individual learning and group work, and to consider this against the requirements of the National Curriculum in science. The framework for science requires teachers to examine traditional teaching techniques within classrooms and laboratories. They need to explore new strategies through projects, investigations, group work, games, simulations, videos, educational drama and role play — and ubiquitous coursework. This

Preface

in turn raises a series of important questions about the nature of individual learning, participation in group work, and the act of communication. For example,

- Can classroom tasks be chosen which allow pupils to demonstrate their understanding?
- Is it possible to encourage pupils to feel that they own the tasks they undertake sufficiently to want to communicate the results to others?
- Within group work, can all the participants be encouraged to own a single task? and
- Can individual contributions be gauged from the outcomes of the group?

The Science in the National Curriculum continues to undergo changes even as we write. We have taken care to use the latest versions whenever we have them. We are pleased that 'communication' retains its place at the head of each programme of study.

If the act of communication could be taken for granted there would be no need for this book at all, so our answers must be yes, yes, yes and yes. The book, then, is our attempt to flesh out the 'how' of communicating in the classroom.

Some of us are good at it, others are appallingly bad. No doubt readers and reviewers will have their opinions of our own efforts within. In writing, we have traded heavily on our communications — written and verbal — with many others and we would like to acknowledge our gratitude for their communing here.

We are very appreciative of friends, colleagues and students in Buckinghamshire (and before that the ILEA), and at Roehampton Institute who have helped. In particular we are grateful to *British Educational Research Journal* for permission to reproduce, in Chapter 2, parts of an article: 'Constructivism in the classroom: enabling conceptual change through word and deed', in *BERJ* 13, 2, 1987; Community Languages Support Services: Hounslow, for permission to use Figures 2.3, 2.4, 2.5 and 2.6 in Chapter 2; Warwickshire Education Department for permission to use Figure 6.8 in Chapter 6; Buckinghamshire Education Department to use Figures 6.5, 6.6, 6.7 and 6.10 and the Association for Science Education for permission to use Figures 4.1 and 4.2 in Chapter 4 and Figures 6.9 and 6.12 in Chapter 6.

We thank Samantha Waite for helping with the typescript.

Di Bentley
Mike Watts
August 1991

Chapter 1

Communicating for Understanding

This book is really about a model of learning, a model that relies almost entirely on acts of communication. Communicating is not exclusive to science of course, though there are many distinctive features about communication in science classrooms. School science does provide us with a very clear context in which to operate, hence the title of the book. We hope that the continued reference to 'science' is not too off-putting for teachers who are interested non-scientists.

The book is also really about groupwork in classrooms, how it operates, how it can be organized and what the benefits can be. Again, groupwork is not exclusive to science, it is often to be found in other parts of school life. In science we have a lot to learn from good practice elsewhere in the system.

We use the word 'communication' quite liberally and recognize that communication can mean many things to many people. 'To communicate', says the dictionary, is to 'pass on, share or exchange ideas'. While it is possible to communicate alone (most of us have little conversations with ourselves from time to time) we are more interested in the communications that take place between individuals, and between groups of people. Groups, that is, bigger than one but of finite size — say, between two and twenty. We are not too interested in mass communication, the general media, political rallies, television or newspapers, semaphore systems or morse code. Our brief is the communication that takes place in school classrooms: within what happens as teachers and children tackle aspects of school science. Beyond that, we are not too fussy — we intend the term 'communication' to cover a wide variety of behaviours and activities in what we take to be schools and science — and to cover the activities of a wide range of people. So, for example, we want to look at pupils communicating with pupils, pupils with teachers, and — in some cases — pupils communicating with adults other than teachers in the school context. In some situations, the term 'adults other than teachers' has earned the acronym AOTs.

The book is based on several premises and we feel it wise to set these out early. First, we work on the assumption that learning is a basic human activity: people learn as a matter of course simply through the act of living and, therefore, of accruing experience. We learn not only through experience, of course, but also through a variety of other modes as well, many of which we discuss in future chapters. Nor is learning just what happens in classrooms with teachers — indeed sometimes learning there can be quite minimal. Learning can take many forms — we learn how to walk, talk, how to be selfish, bigoted, play golf, be gregarious, have good table manners, know our twelve-times tables, understand Newton's Laws and a myriad of other acts in life. Second, learning is inextricably linked with communication. By communication we mean a sharing or exchange of ideas. This is essentially a two-way (or multi-way) process. That is, communication requires some response from the people at the receiving end; communication is *not* a one-way transfer. To have someone communicate *at* you is not communication at all, hence the truth in the kind of expression 'I can't communicate with Chris, it's like talking to a brick wall'.

This is allied to our third major assumption, that both learning and communicating are commonly group and social activities. That is, individual acts are commonly subservient to social requirements and not the other way round. Learning can take place through individual perceptions, conceptualizations, attitudes and actions but the expression of these is usually within a setting where other people are asked to note, comment upon or share some of the meanings involved. Within this we recognize the paradox at the heart of all schooling, as teachers we teach groups but assess individuals. More (much more) of this later.

Our fourth assumption is based on the theory that people interpret things in life in different ways. By 'things in life' we mean common everyday events, phenomena, situations, other people's words, their motives, TV programmes, films, lessons, conversations, drawings, actions and so on. As communication takes place, different people reach different interpretations simply because we *are* all different, even if only in a small way. We all have different life experiences, learn different things, think different thoughts, say different sets of words. This makes the act of communicating more difficult, and so we have to work at it.

Our fifth assumption is that we have a lot to learn from each other. Common practice in one part of the education system is innovatory in another. Well worn techniques in health education or active tutorial work have not yet been assimilated in other parts of school life; a well planned science practical is a model of 'heads-on' and 'hands-on' learning; good primary practice is to be revered by all.

About the Book

Throughout the book we try to tread several careful tight-ropes all at once: to devise a text that is both thoughtful and pragmatic; to cater for many kinds of teachers in a variety of classroom situations; and both adhere to, and move beyond the requirements of the National Curriculum (DES, 1989 and NCC, 1991). While the book is clearly a response to the National Curriculum, it is also about good classroom practice — how to promote discussion, to manage children's questions and feedback, to organize children into groups, to plan and resource talk and writing. We hope each chapter provides a fund of practical plans and procedures. We are as ambivalent as most about the spirit and content of the National Curriculum: we have welcomed its innovation and — at the same time — tired rapidly of its insistent pressure. To readers outside the UK, or those who have been hibernating over the last few years, Britain (England and Wales in particular) now has a centralized curriculum as part of its legislation. This move is one that is happening elsewhere in the world (as described, for example, by Watts and Gilbert, 1989) and encompasses the compulsory years of schooling, from 5 to 16.

Some Terms

Before we move on, though, some terms:

- teachers of science — here we include any teacher for whom some aspect of science falls within their brief, that is, all primary school teachers; members of secondary science departments, teachers of technology in all its forms; geography, and teachers whose role lies outside schools in teachers' centres, advisory posts, university departments, institutes of higher education and so on;
- themes and courses — by 'themes' we mean the usual organizing structure for many primary and junior school teachers, the way they normally plan their topics of work over a period of time. By 'course' we mean the usual units of planning in secondary schools, a 'first year course', the GCSE course for the fourth years, and so on. There may be a large amount of overlap between 'theme', 'course' and 'programme of study' but we do not take this as given.

We do not intend to rehearse all the terminology or structure of the National Curriculum. Throughout, we refer to something called

- 'the Document', by which we mean an amalgam of both the documents Science in the National Curriculum (The Standing Orders) (DES, 1989) and the Science Non-Statutory Guidance (NCC, 1989) and the latest proposals (NCC, 1991). Elsewhere we refer similarly to the 'maths document' and the 'English document' and similar orders and guidelines.
- Activity boxes — which we hope are useful directions towards class activities and exemplars of what is in the text. In addition, Chapter 4 uses a set of

- classroom tasks — where we try to bring together many of the themes that dominate Chapters 1, 2 and 3 in a set of practical tasks.

In summary, then, the remainder of Chapter 1 considers some of the background to the current emphases on communication in the science curriculum. In particular it takes a general constructivist view which equates communication with linguistic and conceptual development.

In Chapter 2 we look at ways in which teachers can create a climate for effective communication in the classroom. This is partly through how they use the physical environment, their own actions and behaviour and their expectations of young people. Chapter 3 follows by dealing with many of the mechanisms for setting up small groups and developing good groupwork practice in the classroom. This is a 'how-to-do-it' chapter — or, at least, 'how at times it has been done well by ourselves and others'. It is not entirely pragmatic since some of our theoretical assumptions keep seeping through. Here, for instance, we expand on the notion that classroom learning can most *effectively* operate by individuals maximizing the operation of the group.

Chapter 4 ties a few knots in a series of arguments. It consists almost entirely of a nine 'Tasks'. These are examples of how many of the themes addressed in the book to that point are brought together. They are studies in the musical sense — practice pieces which incorporate a series of technical moves to illustrate how a number of routines and techniques can be employed in a single, rounded, context. They are also highly practical and form a bridge between the short highlighted activities and the broad arguments embedded in the text.

Chapter 5 tackles what we see to be one of the most difficult of teaching tasks to accomplish — the structuring of learning through open-ended problem-solving tasks. Here, groupwork, communication and individual contributions to group outcomes bear the biggest challenge of all — it is in this sort of context we feel that communicative skills are at a premium.

Youngsters' own evaluation of learning outcomes leads us into Chapter 6 which focuses on assessment issues. In this day and age, there is no communication (or curriculum for that matter) without assessment. We continue here in Chapter 1 with four main issues. First we explore a rationale for communicating in school science. Second, we set out more clearly the model of learning which informs our work. Third, we consider the communicative needs of the National Curriculum and, fourth, we begin to suggest some pragmatic activities for teachers in their work.

School Science

School science has undergone dramatic change in the last few years. Here we want to focus on just some parts of that change. There will always be communication in science classrooms and so, in one sense, the subject matter we discuss in the book is fairly timeless. That said, the communicative acts that take place in science lessons have not always been the central focus of teachers' concern. To some extent that has changed with the introduction of the National Curriculum.

The publication in 1975 of 'A Language for Life' (The Bullock Report) fostered a flurry of activity. There was, of course, interest in issues of language, the readability of text, classroom talk and discussion before 1975, but the Bullock report brought it all to the fore. It raised the profile of all forms of 'language across the curriculum' in most schools and subject areas. Numerous projects were developed and teachers' books written — Barnes, 1976; Martin *et al.*, 1976; Carre, 1981; Sutton, 1981, to note just a few which focused on science. The Association for Science Education, for instance, has always been a strong influence on the direction of science education in the UK, and the ASE developed a line which built on its forward looking documents 'Alternatives for Science Education' (ASE, 1979) and its statement of policy (ASE, 1981). For example, the latter said:

> We are aware that (...) we have not made any firm recommendations as to teaching style and the organisation of teaching and learning. Basically we have argued that teachers and learners should share their perceptions of the educational process, and have therefore placed a high premium on teacher flexibility, the process of self-evaluation, and a substantive plea that pupils should be encouraged to explore their understanding of scientific knowledge and experience through the medium of their own language. The *key* role (our emphasis) for the teacher of science is that of enabling each pupil to relate his or her own perception of scientific understanding to the wider community of scientific ideas. Whilst such a process involves moving the pupil closer to the 'accepted truth', all concerned should accept the absolute legitimacy of the 'perceived truth' — the right of the individual to interpret his or her experience in his or her own words.

This statement was an early declaration of intent which, in some ways, now bears fruit within the National Curriculum.

While these influences had some effect before the National Curriculum, with the publication of the first efforts of the National Curriculum Science Working Party came the first hint of a new, stronger, emphasis. From that point, 'communication' in science was to fall within the purvue of all teachers of science — like it or not.

In those early days of the drafting of the National Curriculum for science (in 1988), science was to be structured in terms of twenty-two attainment targets in four profile components. One of these was called 'Communication' and comprised two attainment targets called 'Reporting and Responding' and

'Using Secondary Sources'. It was intended for pupils right across the age range from 5 to 16, and the report said:

> Central to our view of communication are the notions of range and match. Range in the sense that pupils will, in their reporting and explorations, make use of a variety of styles, media and devices: talking, listening, writing, drawing, modelling, using texts, journals and newspapers, audio and video recorders, data-bases and computer programmes. By match we see pupils varying style and mode of communication to suit audience and purpose; for example, exploring ideas and issues through writing and discussion and, where appropriate, role play and drama, preparing material for 'publication' or presentation to different audiences (for example, teachers, other pupils, and parents), using formal scientific language and symbolism on a variety of occasions, but also utilizing less formal styles and language, where these better serve function and intention. (....) 'Communication' in the context of science studies then, is active, varied, targeted (to audience and purpose), skill-enhancing and central to effective learning and development.

This is a neat expression of our own feelings and approach. It represented, though, a bridge too far for the impressarios of our National Curriculum because, by the final parliamentary orders of 'the Document' a year later (in 1989), much of the spirit and fire had been lost. Communication is no longer a profile component in its own right and is simply 'written through' the attainment targets that remain. While we are resigned to the fact that this is better than nothing at all, and represents a positive move forward from what (cynically) might have been the worst possible case, we certainly rue the loss of what might have been a much better route and higher profile for communication in science.

So, while 'communication' has not quite moved to centre-stage in the statutory requirements of the Science National Curriculum, it has certainly moved onto an apron spotlight. It clearly permeates the attainment targets and the programmes of study in ways impossible to ignore. What teachers might once have thought to be an additional chore, or a touch of luxury, in their science teaching is now there as a matter of legal compulsion.

For example? The summary of the programme of study for Key Stage 3 in 1989 says 'the abilities to communicate, to apply scientific and technological knowledge and ideas are essential elements of a developing experience of science'. Prior to 'the Document' many teachers of science would have thought children's communicative acts to be important, and something to be encouraged within classwork. Now, however, they are seen to be 'essential' and something to be formally assessed within the attainment targets. Moreover, it is the programmes of study which are law — it is these that schools must deliver within the appropriate key stages.

The Communication Gap

One of the major changes in science education brought about by the 1988 Education Act is the part-resolution of a long debate about the accessibility of science. For many years the Association for Science Education (ASE) and the Secondary Science Curriculum Review (SSCR) (among many other bodies) have been arguing that science should be part of the 'minimum entitlement' for every pupil from 5 to 16 (see, for example, ASE, 1981; and SSCR, 1984). Moreover, that science should be as broad, balanced and relevant as possible. In many ways the 1988 Act has brought that about: all pupils must be taught science for the whole of their period of compulsory education. Some pupils (20 per cent or less) may be taught less science, but every pupil will be given the opportunity to participate in a broad and balanced education in science.

This is a very significant change for the processes of communication in school science. First, in making science compulsory, pupils will no longer have the chance to choose only those parts of science which they enjoy the most — or dislike the least. Previously they could opt out of 'difficult' incomprehensible ideas, now they will have to struggle to understand, come to terms with, and even *use* them. Second, compulsion means that the (almost) full ability range of pupils is now expected to do science. Consequently, teachers may well have to re-assess the ways in which they communicate with youngsters, and how they encourage pupils in turn to communicate in science. That is, how to work with the lower-attainer and the less-than-highly-motivated young scientist throughout their school careers.

There is an assumption in all this that all groups will benefit equally if the communication of information and knowledge is increased and made available to all groups in society. Research in communication theory indicates, however, that increasing the total flow of information and knowledge often has the reverse effect: it increases knowledge for certain groups far more than for others. In this way an ever expanding 'information gap' can develop between different groups (see Figure 1.1, from Mcquail and Windahl, 1981, p. 72).

If there is any substance in this, it could have an important implication for science teaching within the National Curriculum. For example, let us assume that the more fortunate in Figure 1.1 are those pupils who, pre-National Curriculum were good at science. They may have opted for science, or may have been those pupils who had access to extra resources at home. In this case, as more information and knowledge of science is made available to them, the greater is their potential for increasing their knowledge and understanding. The 'less fortunate' — those pupils who were unsuccessful — found science difficult, believed themselves incapable of understanding (a very large

Communicating in School Science

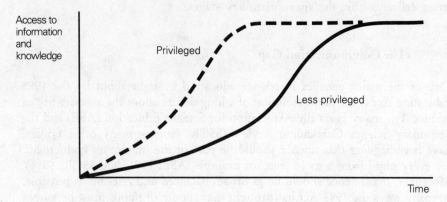

Figure 1.1: Closing information gap, in which the less privileged group 'catches up' with the more privileged one (after Thunberg et al., 1979).

number of girls for example) and who previously opted away from science — cannot capitalize on the increased flow. Like the illiterate caught in an explosion of print, access to more scientific knowledge does not necessarily help. Rather, it can serve to widen the gap between themselves and the 'more fortunate'. This is an important issue: simply making more science available for more pupils may not result in greater scientific literacy and awareness. Science must be tailored to suit the needs of the individual: it cannot be 'O-level physics for all'.

Communicating in Science

The skills of communicating, it might be argued are cross-curricular skills and so there is no need to emphasize the science in the acts of communication. This would be true if scientists, over time, had not become their own worst enemies. Once upon an age, in the days that reached up to the early twentieth century, scientists wrote beautifully, talked elegantly and worked persuasively. As Mermin (1990) points out, some point after that, the art of communicating science suffered a reverse in style so as to reach the point when, he says, the best scientific prose sounds like a 'non-human author addressing a mechanical reader'. He continues:

> The insistence on bland impersonality and the widespread indifference to anything like a display of a unique human author in scientific exposition, have not only transformed the reading of most scientific papers into an act of tedious drudgery, but have also deprived scien-

tists of some powerful tools for enhancing their clarity in communicating matters of great complexity.

Somehow scientists, possibly because of the recent explosion and the resultant huge quantity and detail in scientific research, have contrived to hide controversy, anger, irritation, pleasure, ecstasy behind an authoritative grey surety.

So what does this have to do with school science? Well, science need not be this way. That is, there is a sense in which science once again needs to communicate its humanity, and scientists' their glee, as the business of science goes through its everyday cycles. There is no doubt that behind the scenes science is as tumultuous, divergent and full of conflicting voices (see Gilbert and Mulkay (1984) 'Opening Pandora's Box', and, more lately, Holt's (1990) work) as ever. This is clearly a public issue as scientists face television audiences and contradict each other on the nature of 'mad-cow disease', 'lysteria hysteria', 'the greenhouse effect and the ozone layer', 'food irradiation' et cetera, and where major law cases and possible gruesome miscarriages of justice hang on the threads of forensic science.

Language in Science

We mentioned earlier some important work which has already provided us with a backdrop against which to discuss communication in school science. Much of this has taken the form of 'language studies' and has considered the nature and use of scientific language in the school laboratory. This work is important here on two counts:

- it allows us to begin discussion of communication between teacher and taught; and
- it provides an entrée to the particular views we have on learning and school science.

Through both text books and teachers, school science has inherited much of the 'bland impersonality' and lack of clarity Mermin regrets. This in turn has exacerbated a problem: many of the difficulties experienced by youngsters in science classes arise from both the technical language used and the non-technical but supportive language that surrounds it. Let us begin with the technical.

A technical term in science is one for which we (as scientists) reserve a fairly guarded definition and set of implications. Some very common examples from physics are 'energy', 'force', 'work', 'pressure', 'power'; from biology: 'living', 'animal', 'cell', 'reproduce'; from chemistry: 'reaction', 'bond', 'matter', 'solution', 'suspension' and so on. These may, at first glance, seem very un-technical terms — words like electromagnetism, photosynthesis and polymerization seem much more redolent of school science. However, we have chosen the first list precisely because they have dual (at least) meanings in both science and everyday life. There is a wealth of research which has explored the difficulties youngsters experience when trying to 'come to terms' with the words of science — Kingdon and Critchley (1982), Watts and Gilbert (1983), Wellington (1983), Bell and Fryberg (1985) are but a few. Driver, Watts *et al.* (1990) provide a very comprehensive bibliography of the field. It is commonly the case that teachers of science make strenuous efforts to define technical terms as soon as the class is likely to encounter them. This is not always successful — Richards' (1979) work is still a cogent reminder of the complex language registers used in science education:

> Characteristically such language shows increased formality of style, makes use of a specialised vocabulary and contains certain repetitive patterns of syntactic structure which are often more complex than would occur in the language normally employed by the pupil.... Specialised language forms appear more frequently in science subjects than in arts subjects, while increased formality and language outside the pupils' range are more evenly distributed throughout the subjects. In this respect biology clusters with physics, mathematics and chemistry, although the features are more strongly represented in physics and chemistry.

In short, this simply means that youngsters are prone to confusion and misinterpretation when deciding on what, exactly, is force, energy, radiation and so on.

And if the technical terms present problems, hold hard! There are also the non-technical words. Marshall, Gilmour and Lewis (1990) are but some of the latest to point out that although pupils may appear to understand the words being used, it is often the case that their understanding is not congruent with that of their teachers. Over a decade earlier, Cassels and Johnstone (1980) noted that

> the problem lay, not so much in the technical language of science, but in the vocabulary and usage of normal English in a science context. Pupils and teachers saw familiar words and phrases which both 'understood', but the assumption that both understandings were identical was just not tenable.

Clearly there has not been much change taking place. What sort of words are we talking about? Well: accumulate; agent; consistent; contribute; device; component; control; initial; source; crude; devise; estimate; diagnose; evacuate; exert; influence; limit; random, are but a few to have been explored. Cassels and Johnstones' (1985) list is much longer.

Towards Understanding

At this stage there is a sense of 'so what?' in the air. So what? We *know* that youngsters do not fully understand science — why else would we be teaching them? We know that only the brightest succeed — the ones that can get to grips with the language of science. We know that language is a great discriminator — we only have time to get through the syllabus/programmes of study without having to worry about teaching English too. And so on.

Were it only that easy. Our stance at this point is to shift towards our constructivist philosophy: 'How we look at the world depends on the concepts we know and use in order to understand it' (Howard, 1987).

In sharing the philosophy of constructivism, we subscribe to an 'ism' that seems to be growing in influence within science education. We describe ourselves as constructivists, and part of a way of thinking about learning in science.

Constructivism

Constructivism is a broad term which covers a range of theories — and theorists — who share common points of view. Mahoney (1988), for instance, notes that:

> Constructivism refers to a family of theories that share the assertion that human knowledge and experience entail the (pro)active participation of the individual.

For instance, in Vygotsky's (1986) work he describes two kinds of concepts — 'spontaneous' and 'scientific'. He suggests that children build spontaneous concepts long before they are conscious enough of them to be able to define them in words. They do this through their everyday interaction with people and objects in the world. In contrast a scientific (non-spontaneous) concept is commonly formed through a verbal definition and its use in non-spontaneous circumstances. That is, it starts life in the child's thinking in a way which spontaneous concepts only reach much later:

> In working its slow way upwards, an everyday concept clears a path for the scientific and its downward development. It creates a series of structures necessary for the evolution of a concept's more primitive, elementary aspects, which give it body and vitality. Scientific concepts in turn supply structures for the upward development of the child's spontaneous concepts towards consciousness and use.

Vygotsky built his work upon a large body of empirical investigation and it gives a firm basis for the emphasis on the ability to express concepts (both spontaneous and scientific), as a vital part of school life. In the complex relationship between thought and language Vygotsky regards word-*meanings* as a phenomena of thinking, of dynamic formations which evolve and change as the child develops. Words are not merely used to express thought, they are the means by which thought comes into existence. Without words it would not be possible to communicate to someone else the limits of a concept.

In other parts of our own work, we have tended to focus on another member of the family — through Kelly's (1955) Theory of Personal Constructs. We have discussed the impact of Kelly's work on education elsewhere (Pope and Watts, 1988; Watts and Pope, 1989). In his theory, Kelly takes his root metaphor as 'man-the-scientist'. He invites us to suppose that, in the course of their everyday lives, everyday people act like everyday scientists. The essence of Kelly's ideas is that each person builds a model of the world which enables him/her to chart a course of behaviour within it. Like a scientist's model, this model is subject to change over time since constructions of reality are constantly being tested and modified to allow a better working model to be erected. 'Better' or 'worse' here means how well it serves people in predicting and sorting out what is happening as they go about their daily business.

Human beings are individuals and they are also highly social animals. No individual is isolated from group interaction and so living — and learning — takes place within a powerful social milieu. The central point here is that working out thoughts and ideas with yourself, or exploring issues with friends or colleagues, is vital for learning and for communicating. Shaping new concepts, changing old ones, forming new links and connections or severing outdated ones, happens as we communicate with ourselves and when we share ideas (communicate) with others. A good way to illustrate this is to meet Cathy and Colin previously noted in Watts (1983) and Watts (1991).

Cathy

Cathy is 14 years old and, shortly after a science lesson, is talking about the nature of heat. She is a very quiet yet composed young woman who, in class, may not volunteer ideas but has well-formed opinions. Within her concept of heat, she makes a distinction between heat and 'cold', the latter being something quite tangible: her mother often has to tell her to keep doors closed in winter so as not to 'let in the cold'. She continues her discussion by making the commonplace observation that 'heat rises'. As she talks she develops a model of the earth's atmosphere where the temperature gradient is such that the temperature increases with distance from the surface of the earth. That is, 'it gets hotter as you go up in the air; it just gets hotter and hotter'.

Does it continue so, for ever into space? No, there is a limit, she says. The upper edge, the layer between the atmosphere and space, is a fairly firm boundary. Space craft have to break out of the earth's atmosphere, she knows, because they need an 'escape velocity' for that to occur. The film that retains the atmosphere around the earth is, she thinks, made of ozone and now has some holes in it — quite possibly allowing air to escape into space. The layer is 'a kind of skin that keeps the air in'. What is the temperature like at this level?

> It is very very hot. That's why things like meteorites burn up when they come into the atmosphere. That's why the space shuttle needed those heat resistant tiles on it.

Mountains are high, why is it possible for snow to remain on high mountains? A few seconds' pause and puzzled thought, before:

> No, I'm not sure. But I do know its hot up there ... my friend went skiing at Christmas and she got sun-burnt.

Colin

Colin on the other hand is talking about light. Like Cathy he is also 14 and, as does Cathy, attends a local London comprehensive school. He is a chirpy character, full of ideas, though something of a reluctant scientist in that he avoids science lessons when he can because often

> they're boring'. 'Light' he says 'is funny stuff. I'm not really sure how it works ... I know it travels very fast — well at the speed of light really! But I don't understand how we can SEE it, if it travels that fast.

His discussion of the properties of light focus first on its speed. He describes a slide projector that is projecting a coloured slide onto a screen (see Figure 1.2). Colin points out that the light is generated by the bulb in the projector. It then travels across to the screen where the light stops. The screen, he says, must act like a barrier:

> The light must stop there, stop completely ... Otherwise we'd see the picture moving, wouldn't we? Sort of shimmering.

He uses a similar rationale later, when discussing the transmission of television programmes. He is aware that there is a large transmitter in his neighbourhood which is responsible for transmitting some of the television programmes he watches. He is also aware that television transmission use electromagnetic waves which travel at the speed of light.

Communicating in School Science

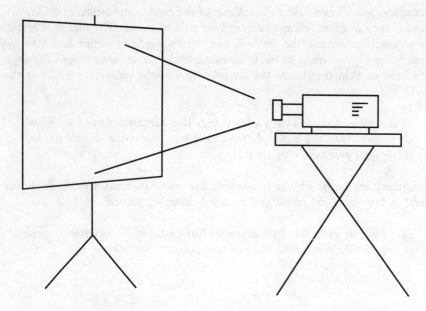

Figure 1.2: Colin's description of a slide projector

The programmes leave the transmitter and they must travel at the speed of light through the air to the aerial on the house. Then they must come down the wire from the aerial to the television set. And the television must have a special mechanism in it to slow them all down, because when you watch them, the programmes are all back at the normal speed, aren't they? ... I mean the people walk around and talk in the usual way, don't they?

His ideas about light were not without awe and wonder.

You know, he asked, when you are sitting indoors on a sunny day and the curtains in the room are just parted so that you can see beams of light coming into the room? Well, you can see through the beams to the other side of the room, can't you?.... What I don't understand is how we use light to SEE, but can see THROUGH it at the same time.

Alternative Frameworks

Cathy and Colin have particular ways of looking at the world. Their ideas are not just isolated flights of fancy, but are imaginative ways in which they have attempted to explain how things work. They have remembered parts of the science they have been taught in class but it is highly unlikely that this is exactly what they were taught. Somewhere between the classroom and the discussion afterwards they have re-interpreted issues, pieced together some parts from previous lessons, added some ideas and questions from personal experience and constructed a theory for themselves that makes sense and explains the situations at hand.

Their theories about light and heat are 'alternative' to orthodox physics in ways too numerous to chart in full. Physics, for example, recognizes heat only as a verb — the process of transferring energy from regions of higher to lower temperatures which are in thermal juxtaposition. There is no such thing as 'cold' (simply lower temperatures); it is not 'heat' but hot air that rises; temperature decreases upwards through the atmosphere — a notion that is used to explain the planet's weather systems; there is no 'skin', the atmosphere becomes gradually less dense to the point where it simply becomes 'space'; meteorites burn because of the exchange of energy through friction as they encounter the earth's atmosphere; and so on.

Colin's conceptions of light are very common — it is a popular idea that light does not necessarily have to bounce off an object (like a screen), but can form 'pools of light' that observers simply look at. Physics would describe light differently — we see because light comes directly to the eye from a source, or because it reflects off objects. This is the only way we can see things. Televisions do not 'slow down' the transmission, here Colin is mixing the medium with the message. And the superposition of waves is a fundamental property of light — that waves can pass right through each other without hindrance. It is a basic principle in physics.

Personal theories like Cathy and Colin's are not simply the result of misconstruing words, technical or otherwise. A misunderstanding of terms can contribute to an unorthodox world view, but constructivism regards such personal theories as a normal part of everyday life in their own right. They are the way we make sense of the world, how we go about learning and communicating ideas. We construct models for all sorts of situations to allow us to structure them for ourselves. Theories like this are not necessarily well thought out, and a person may well have several sets of ideas at the same time which seem to conflict with each other. Coghill (1978), for instance, describes a youngster's different meanings for the terms 'odd' and 'even'. In working alongside a young girl she came across the two words in a mathematics book. The question was:

1, 3, 5, and 7 are odd numbers. What is the next odd number?

The girl clearly did not understand. Coghill says:

To help her understand, I told her that to find out if a number was an odd number or an even number, we could take that number of things,

> share them between two people and if each person has the same amount it was an even number, and if there was one left over it was odd.

They went on to try this with several sets of things, books, pencils, bricks, etc. and the girl seemed to grasp the idea quickly. A little later in the maths book there came the question:

> What is the number of your house? Is it an odd or even number?

The girl said:

> My house is number 15 and it's even. We share it with the people upstairs.

Since the 1970s, a growing body of research in science education has focused on pupils' knowledge and understanding. In the literature these have been called 'alternative conceptions', 'alternative frameworks' or pupils' lay theories. They are alternative to the orthodoxy of science.

The field is now wide and substantial, and comprehensive reviews and discussions of this work appear, for example, in Gilbert and Watts (1983), Driver *et al.* (1985), Osborne and Freyberg (1985), Fensham (1988), White (1988), Adey *et al.* (1989). Excellent studies of lay theories in many other walks of life are to be found in Furnham (1988) and Semin and Gergen (1990). The Children's Learning in Science Projects (CLIS) based at the University of Leeds has been building a wealth of research and documentation in the field for almost a decade (Driver, Watts *et al.*, 1990).

Within education more generally there has been considerable research on the approaches young people take to learning. Within this, studies of cognition have usually followed one of two central directions. The first, as in the bulk of Piaget's work, emphasizes the skills, processes or logical 'structures' thought to describe children's intellectual operations. The second explores the content of young peoples' ideas and understandings and has provided a fairly extensive mapping of the content of youngsters' theories, conceptions or alternative frameworks. It gives a good indication of the ways youngsters conceptualize parts of science.

Some alternative conceptions appear at all levels of schooling from infants to A-level and beyond, to undergraduate and postgraduate level. This is taken as an indication that such conceptions can be fairly resistant to change. The bulk of the research makes it clear that school pupils have:

- a number of unorthodox ideas and understandings about a wide range of topics;
- ideas that can remain intact in the face of normal everyday teaching;

- understandings that shape how they make sense of new data and information,
- and which can even persist in the face of seemingly strong counter argument and evidence.

The topics investigated have focused on concepts across the sciences — from 'heat', 'energy' and 'light' in physics to 'plant nutrition' in biology, the 'nature of matter' in chemistry and 'the place of the Earth in the universe' in astronomy.

Constructivism, then, holds that a person's overall system is comprised of a collection of distinctive but related clusters of concepts or constructs. So, for example, a person's constructions of issues in education may very well be different from their construction of legal matters. The two though, may be related, linked through ideas of — say — social justice, the exercise of power, the role of punishment, or teachers' responsibilities *in loco parentis* and so on.

Neither constructivism in general nor the theory of Personal Constructs in particular is a suggestion that all people are actually scientists, simply that viewing people in their 'science-like aspects' can illuminate human behaviour. It is very similar in spirit to the proposals of National Curriculum Science Working Group (DES, 1988) who adopted a very clear 'child centred' perspective as a broad rationale for their work. The child, they said, is the agent of his or her own learning in science. Children's learning in science is linked by analogy to scientists' advancement of ideas, hypotheses and principles when faced with new phenomena. Their prior knowledge and initial theorizing are therefore important as part of the process of reaching a scientific understanding of the world around them.

Their words are used by the National Curriculum Council when they say (NCC, 1989):

> Viewed from this perspective, it is important that we should take a pupil's initial ideas seriously so as to ensure that any change or development of these ideas, and the supporting evidence for them, make sense and, in this way, become 'owned' by the pupil. The ideas of young children can be essentially scientific in so far as they fit the available evidence even though they will (...) fall a long way short of, or even be inconsistent with, formal theories.

Communication

Whilst, then, we are always communicating, we are sometimes more effective in our communications than at others. But we keep on trying. We noted earlier that communication can take place in many ways: thought, talk, discussion, drawing, painting, singing, drama, number, posture, music, dancing and many more. Often we do not have to say anything at all to communicate — teachers often communicate through non-verbal means. A familiar situation, for instance, is recalled by one girl (in Bentley, 1987) who tells of her response to the headteacher and compares it to that of a rabbit transfixed:

> When he looks at me, you know, like he does in assembly, he kind of glares and doesn't say anything. Then you know it is time to sit down and be silent. He just stands there looking at me and I feel like a rabbit in the headlights of a car. I'm kind of paralysed.

In the book we focus on both verbal and non-verbal means of communicating, the 'bread and butter' of communication in classrooms. Our point of departure is to link these forms of communication with conceptual development. As we tackle these issues we make reference to three kinds of communicating we call 'explication', 'expectation' and 'exploration'.

Explication

The explication of ideas entails all the processes of making them available to oneself and to others. That is, providing all the opportunities for both youngsters and adults to find out what exactly it is that they *do* think for themselves. It is the process of putting thoughts into words, of making explicit what might — up to then — have been implicit or tacit. It can be a personal process or a communal one. In classrooms, it certainly has to be set within the kind of atmosphere that will not inhibit clear thinking and personal disclosures, something we describe before as a 'non-threatening environment'.

Expectation

By expectation we mean all the processes of testing ideas and hypotheses, working out the expectations and implications of holding those sorts of opinions. Personal ideas can range from the speculative to ingrained, the temporary tentative to the fixed and final. Testing these ideas means setting them out for comment and criticism, checking them against other ideas and their fit with experience — generally 'holding them up to the light'. If the idea says X, what are the implications, how can these be understood, and what are X's limits? Again this can be a solitary or a corporate activity and also requires a fairly trusting environment for it to happen without people being hurt or inhibited.

Exploration

This process is one of trying out other people's ideas for size, testing one's own constructs and conceptions by temporarily adopting other points of view, 'putting yourself in someone else's shoes': what Kelly calls exploring the 'what if?' It is commonly the case, as Gauld (1989) points out, that these processes can lead to youngsters reconstructing their ideas and how they need to continue over time to be effective:

> The evidence presented demonstrates the importance of such activities but also the limitations which occur when these activities are confined to the brief period during which the topic is dealt with in school.

We return to these three forms on occasions as we explore different aspects of communication.

What the 'National Curriculum Science Document' Currently Says

The term 'Communication' itself appears throughout the wording of the National Curriculum documents and appears in the text, too, in different guises. Here we paraphrase the main sections which come at the start of each programme of study for the key stages:

Key Stage 1

Through their study of science, children should use a variety of communication skills and techniques involved in obtaining, presenting and responding to information. They should also have the opportunity to express their findings and ideas to others and their teacher, orally and also through drawings, simple charts, models, actions and the written word. They should also be encouraged to respond to their teacher and to the reports and ideas of other pupils and become involved in group activities. In order to supplement their first-hand experiences they should be introduced to books, pictures, videos and to the use of computers.

Key Stage 2

Children should have opportunities to continue to develop and use communication skills in presenting their ideas and in reporting their results to a range of audiences, including children, teachers, parents and other adults. In giving an account, either orally or in written form, they should be encouraged to present information in an ordered manner. They should be introduced to the conventions involved in using diagrams, tables, charts, graphs, symbols and models. Children should be given opportunities to participate in small

...scussion and they should be introduced to a limited range of books, ...nd other sources from which they can gain information. Children ...se the computer to store, retrieve and present their work.

Key Stage 3

Pupils should be given the opportunity to extend their use of scientific and mathematical conventions and symbols. They should be encouraged and helped to read actively and for a purpose, through the use of an extended range of secondary sources. They should take increasing responsibility for selecting the resources on which they draw. Pupils should be encouraged to express their ideas by various means and to respond to those of others; and to record their work. They should begin to use, with increasing confidence, information and data accessed from a computer.

Key Stage 4

Pupils should be given the opportunities further to develop their skills of reporting and recording. They should be encouraged to articulate their own ideas and work independently or contribute to group efforts. They should develop research skills through selecting and using reference materials and through gathering and organizing information from a number of sources and perspectives. They should have the opportunity to translate information from one form to another to suit audience and purpose and to use data-bases and spread-sheets in their work.

In both the 1989 and 1991 proposals, these words represent a formal requirement and a specification of what communication is to be in science. There are overlaps, of course with what is expected in the other two core subjects, English and Mathematics, and we consider that a little later. It is first worth looking at the 'communicative acts' set out in the New Attainments Targets for science.

New Attainment Target 1

New Attainment Target 1 is a major part of the work to be done in science and needs special consideration. It is all about the investigative, exploratory and communicative skills that are part of science. The weightings associated with this component — compared to knowledge and understanding — varies with the key stage, and means that it represents about half of all the science work in science at primary level (5–11 years) and about a third at secondary level (11–16 years). This is a considerable amount of work. And, although NAT1 is to be assessed and reported separately as a profile component in its own right, it is not to be taught separately. The intention is that the investigative and exploratory work is to be integrated throughout the Programmes of Study. The wording of this Attainment Target is

> Pupils should develop the intellectual and practical skills that allow them to explore the world of science and to develop a fuller understanding of scientific phenomena and the nature of the theories explaining these, and the procedures of scientific exploration and investigation. This work should take place in the context of activities that require a progressively more systematic and quantified approach, which draws upon an increasing knowledge and understanding of science. The activities should encourage the ability to:
>
> (i) plan, hypothesize and predict
> (ii) design and carry out investigations
> (iii) interpret results and findings
> (iv) draw inferences

The 'Communicative' Descriptors in the Old Attainment Targets (OATs)

It is worth looking through the first document (DES, 1989) to give some indication of how 'communication' was seen to happen. First, there are the more straightforward descriptors to recognize, where 'the 1989 Document' uses terms such as:

children should be able to

support their views (OAT2: be able to support their view about environmental issues concerned with the use of fertilisers in agriculture and horticulture, based on their practical experience);

organize information (OAT2: be able to organize information from a number of sources to present an understanding of the relationships between population growth and decline and environmental resources);

describe (OAT3: be able to describe the main stages of the human life-cycle);

give an explanation (OAT3: be able to give a basic explanation and evaluation of the impact of life supporting technology in improving and sustaining the quality of life);

keep a diary (OAT5: be able to keep a diary in a variety of forms of change over time);

argue for and against (OAT5: be able to argue for and against particular planning proposals in the locality which may have an environmental impact);

select and weigh evidence (OAT5: be able to select and weigh evidence to form reasoned judgments about some of the major ecological issues facing society);

give an account (OAT7: be able to give an account of the various techniques for separating and purifying mixtures);

state (OAT9: be able to state qualitatively the relationship between pressure and winds);

evaluate (OAT10: be able to evaluate the design of a structure or artefact, by balancing considerations of strength, choice of materials and cost);

express (OAT11: be able to express the relationships between the following quantities in the appropriate units: charge, current, voltage, resistance and electrical power);

discuss (OAT17: be able to discuss clearly with others their way of thinking about some experiment which is new for them).

These terms are taken from a variety of levels within the old attainment targets and so relate to different levels of communicative sophistication. We can but rue the loss of this sense of expression. The most recent proposals (NCC, 1991) give few illustrations bar 'describe', 'observe' and 'explain'. Once again an opportunity has been lost to enrich the science curriculum.

Second, though, there are many other types of 'operational descriptors' within the Document which are also 'acts of communication'. What do we mean? Well, throughout the levels of attainment the NCC use such words as pupils should 'be able to understand', 'know that', 'understand how' 'relate' and so on. There is no reason at all why these processes should be personal, private and internal acts of understanding, knowing, of the 'conversations in the head' type. Indeed, there are many reasons why we should encourage youngsters to share their knowledge, understandings, observations and evaluations. Learning can be enhanced where such issues are concerned — learning can be open, cooperative and still highly effective.

OAT12 and OAT17

These two attainment targets were special in two ways:

Communicating for Understanding

they were a departure from traditional science as taught before the National Curriculum;

they both feature 'communicative acts' in a particular way.

Old Attainment Target 12, concerned with information technology and electronics, represented the 'hardware' elements of communication (in some circles) and so deserved a separate note. Attainment Target 17 focuses on 'the Nature of Science' and relates to the notion of science as a social activity and the programme of study leaned heavily on discursive methods to realize its aims. Both are now incorporated into the body of the curriculum which gives rise to the point of this book: communication is 'writ through' the curriculum and not just one small 'discursive' section.

Progress and Progression in Communication

Although some of the terms (words like 'describe') are used throughout the attainment targets, it is clear that they are intended to mean different things at different times. That is, the kind of description a child provides at level 3 will be (and is expected to be) different from that given by a young person at level 8. What progression can we expect? It is easy to see that a distinction can be made between the quantity and the quality of learning that can take place across any period of time. It is easy, too, to appreciate that assessing progress in terms of quantity involves checking to see how many points a pupil might remember, how well she or he can solve problems, how much their vocabulary has expanded and so on. Assessing quality is not so straightforward. Consider the following example from school physics, one we use again in chapter 3. Pupils are asked to say what forces they think might be acting on the astronaut and the spanner in Figure 1.3 below (Watts and Gilbert, 1986). The responses range from:

1 'Dunno'.
2 'The astronaut is weightless so there's nothing'.
3 'There's no gravity on the moon because there is no air or atmosphere, so there are no forces at all'.
4 'Gravity on the moon is less than that on the earth and it works upwards. The astronaut might stay down because he has heavy boots on, but the spanner will float up'.
5 'There's the force of gravity down and a reaction from the moon's surface up, which are equal and opposite. It is the same for the spanner but without the reaction and so it will fall'.
6 'There are far too many forces to be able to discuss them all. There are electro-magnetic forces internal to all of the objects on

Which drawing do you think best shows the GRAVITY on the spanner?

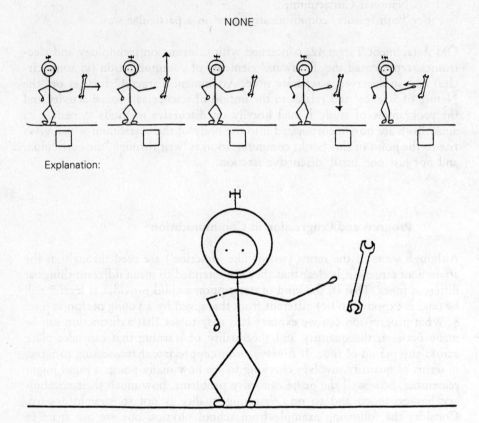

Explanation:

Figure 1.3: Sample of school physics question on the effect of gravity

view, weak and strong, frictional — possible electrostatic — and gravitational forces between them, and so on'.

The responses certainly increase in length and the number of points that are made, so some judgment could be reached as to the 'quantity' of response. Most readers, though, would also recognize that there is a different quality about them. It is not that they are more correct in scientific terms (all of them are incorrect physics in some form or another) but that the language changes, the complexity grows, the display of understanding is better. The odd one out, of course, is the first one — since we cannot judge whether it reflects the response of someone who has few — if any — ideas at all, or a potential level

Communicating for Understanding

10 student stunned for a moment by the complexity of the question (or in the middle of an off-day).

Certainly, as age, stage, and schooling wear on from 5 to 16, most teachers would want to see youngsters make progress in their abilities to communicate. In this sense we need to examine the National Curriculum very carefully for the signs and signals for what communicative ability entails so that we can provide the opportunities for pupils' skills to grow. We return in later chapters to the problems of progression and assessment.

A Checklist of Communicative Skills

To end this chapter we pick up two threads that are more fully developed as the book progresses: the nature of communicative skills; and the assessment of skills. This first list serves as an overall check for teaching at any phase or stage. It can be used both to note how the opportunity to learn has been provided — and also that learning has actually taken place. Activity 1.1 asks the reader to judge if:

ACTIVITY 1.1

Pupils can (have they been given the opportunity to)
1. respond to direct questions?
2. listen to and act on straightforward instructions given orally, in writing or diagrammatically?
3. give an account of an activity or investigation?
4. appropriately sequence the major aspects of an account or set of instructions?
5. use diagrams and other conventions as appropriate?
6. set out their own ideas and thinking on an event or issue?
7. prepare an account for an agreed audience?
8. contribute to group discussions?
9. use relevant scientific language and notation?
10. use a variety of textual material?
11. match responses to audience and purpose?
12. select appropriate sources, and organize information from a range of sources?
13. translate information from one form into another?
14. explore points of view other than their own?

Within each of these skills in the list there are clearly *degrees* of skilfulness — and the assessment of pupils' skills at the appropriate level is obviously important. Consequently, the skills involved in, say, describing an experiment will depend on the nature of the task, the context in which it is set and the possible routes to an outcome. Trying to assess whether or not skills have been deployed successfully will depend on their role in developing the investigation.

Any one of the skills above can be spread over a spectrum of performance — on, say, a five point scale for assessment (or self-assessment) purposes. So, for example, 'talking' might be judged in the following way:

ACTIVITY 1.2

1. is able to give some basic information, relate personal experience and opinion simply, with some hesitation and repetition, when asked.
2. can give basic information and relate personal experience and opinion clearly with some sense of audience.
3. volunteers information clearly. When asked, can be articulate about personal experience and opinion, adapting presentation with a clear sense of context and audience.
4. volunteers complex information in a clear and organized way. Can participate well in group discussions and empathize with other points of view.
5. can initiate sustained discussion and maintain a logical or emotive argument. Can coordinate the discussions of others and synthesize group ideas and debate.

However, in the context of our aims for this chapter, assessment is something of a digression and is tackled in full detail later.

Chapter 2

Creating a Climate for Communication

This chapter is about how we shape the learning environment to aid communication. It follows a thread of argument that can be summarized as follows:

a our major concern is enabling youngsters' conceptual change;
b current theories of conceptual change require considerable self-exposure by youngsters of their existing ideas. We have described this exposure in terms of the explication, expectation and exploration of youngsters' conceptions;
c willingness to articulate and explore personal theories is dependent upon the learning environment prevailing in a classroom;
d the learning environment is created both by what a teacher says and the way that she/he behaves;
e youngsters are shrewd judges of behaviours which indicate trust, sympathy and empathy in teachers and prefer to work for and learn with teachers they trust;
f effective conceptual change requires that teachers become skilled in both the verbal and the non-verbal cues they initiate.

The argument is an attempt to bring together two distinctive and often separate sides of communication: on the one hand language and language development, and non-verbal actions on the other. Within the chapter we explore aspects of what we call a 'non-threatening learning environment' (NTLE). At first glance it might seem a mildly eccentric inconsequence — rare are they who champion a *threatening* learning environment. The caricature of the towering, bullish, PE teacher; the piercing martinet of an old fashioned primary school teacher, the ex-army boxer deputy head are — thankfully — fading stereotypes.

A NTLE seems to us, though, a notion worth developing on two counts. The first relates to communicative acts within the psychosocial atmospheres of school classrooms, whilst the second stems from the kind of research work and curriculum development we describe in other chapters.

Our experiences foster the conviction that, on the whole, life in school classrooms is conducted in a fairly robust atmosphere, redolent of the normal chiding, teasing and banter that occurs within pupil peer-group and teacher-pupil interactions. On the heavy-handed side, sarcasm, abrasive wit and verbal bullying are not uncommon tools in the armoury of the hard pressed teacher — or youngster. The model of the dour, humourless, disengaged authoritarian is sometimes offered as an appropriate role for initiate teachers at the outset of their careers — the old adage 'never smile before Easter' sometimes being proffered as having 'a germ of good psychological sense in it' (Marland, 1975). Needless to say, warmth, humour and sensitivity are also part of teachers' approach, though arguably perhaps not always with the same regularity and consistency as the caricatures above.

An interesting piece of research recently has drawn attention to the need to create a 'collaborative learning environment' (Hart, 1989) in classrooms. Such an environment, Hart argues, provides a framework for whole class work and formal and informal small group work. Hart's descriptions are not dissimilar from our own thoughts on a non-threatening learning environment. For example, Hart suggests that to bring about collaboration between members of the class, the environment needs:

> the existence of relationships of trust between children ... is a crucially determining factor ... children will not take risks or choose to collaborate with others if they think other children will make fun of them or disrupt their work. The fostering of a safe environment in which rights, responsibilities and the need for mutual respect are explicitly negotiated with the children, and, if necessary, enforced by the teacher is therefore essential. (Hart, 1989 p. 12)

Clearly, the atmosphere of classroom is of importance for learning and the teacher contributes to creating this by a variety of communicative means. However, not all communication is verbal: non-verbal actions, postures and expressions abound in the normal day-to-day discourse of school life. Frequently, much of a teacher's organization and communication within the classroom is geared to the pursuit of social control. For example, many teachers manipulate class dynamics for this purpose. Sitting the pupils in rows of desks to respond only to the person with whom everyone has eye contact — the teacher — is one way of ensuring that communication between pupils is lessened. The social dynamics being emphasized here are 'compliance' and teacher control. Talk is channelled through the teacher and the conditions discourage interaction between class members — a far cry from Hart's collaboration and a non-threatening learning environment! Other similar examples abound. Marland, for example, goes on to advise the beginning teacher to, whenever possible, influence the misbehaving pupil, or the pupil about to misbehave, silently and without the rest of the class knowing:

> Perhaps a small gesture will catch his eye and he will look up. Then a mere continuation of the stare may be sufficient, but it can be strengthened by a frown, or even a smile. This last may sound surprising, but a smile indicates you know the pupil was up to something he shouldn't have been, that you are not furious — yet — and that if he stops all will be well. A mouthed but soundless word or two can also be added occasionally. Such tactics avoid advertising the unsuitable behaviour to other pupils, and avoid the attendant risk of encouraging others to join in. This even creates a conspiratorial feeling between the teacher and the would-be wrong-doer that leaves a pleasurable rather than a thwarted feeling in the pupil.

Argyle (1969) has argued that as much as 60 per cent of the information gained in *any* communication between two individuals is gleaned from the non-verbal aspects of that communication. Feldman and Orchowsky (1982) report the use of non-verbal behaviours in teaching contexts. They add a further gloss to Argyle's figure:

> That successful students received more positive non-verbal behaviour than unsuccessful students seems quite reasonable ... the *degree* of non-verbal positiveness of the teacher could have potentially important effects, not only on student learning but upon their self concept as well. (Our emphasis)

That is, differences in teachers' non-verbal behaviour might well facilitate learning because they enhance the clarity of the feedback given to youngsters. In our terms, actions which enable youngsters to feel positive about themselves aid the processes of cognitive change. But one might well ask, what are these 'positive non-verbal behaviours' that are so important? Richey and Richey (1982) and Bentley (1983), for example, have identified frequent eye contact, frequent hand gestures, open postures and physical closeness to be behaviours which indicate psychological acceptance.

The learning environment (non-threatening or otherwise), then, rests as much — if not more — on the way a teacher acts or 'speaks' through organization in the classroom as it does upon what he or she might say. It also rests upon the ability of the other participants in the room to decode the teacher's intended meanings. Yeomans (1988) has an interesting analysis of how primary teachers organize the learning environment to help pupils feel secure when working as a whole class. He identifies four aspects which help to establish a collaborative atmosphere for pupils working together.

1 Arranging the physical environment to create a small group impression, by creating different areas in the room and moving away from the 'backs of heads' features of large groups.
2 Concentrating on developing a warm one-to-one relationship with all individuals.
3 Treating the whole class in a family way — gathering round for a story at ritualized times.
4 Creating tasks and opportunities when children are deliberately encouraged to collaborate in groups — producing a group report.

The Physical Environment

The reference to the physical environment in Yeoman's four aspects is an interesting one. Generally such environments are referred to as 'classrooms'. By 'classroom' we mean any room or location in which (science) teaching occurs. Unlike other countries, in the UK it is common for secondary teachers to stay put and pupils to move from room to room for different lessons, changing classroom some three, four or five times a day. Largely, teachers and students must work in conditions which neither have planned or furnished, but which in many ways both have a real stake in making as hospitable as possible. As we move around schools we are constantly struck by the huge efforts made to embolden the surroundings — from the mundanity of clean boards and tidy cupboards to dazzling wall displays, flowers and plants, interactive displays and exhibitions, and children's work in abundance. In contrast, the best that can be said of some areas in some other schools is that they can be 'cleared for action' and be businesslike in a short time.

The whole business of a stimulating and challenging learning environment is an interesting one. What makes a particular classroom stimulating whilst others are less so? Display certainly plays a part. HMI have criteria which focus on the use made of display in schools, as do many local inspectors. There is no doubt that many primary schools create an immediate impression of being 'good schools' because of the high quality of their displays.

What are the criteria for a stimulating learning environment? They are clearly different for different ages of pupil. What is stimulating for 7 year-olds is not what stimulates 17 year-olds. In this section we explore what they might be for all sections of the education system.

Aspects in common across all ages.

Does the display:

- Value and celebrate the achievements of all pupils whatever their ability?

Creating a Climate for Communication

- Extend the aspects of the curriculum being taught in that room at the moment?
- Stimulate and challenge further thinking for pupils?
- Promote positive images of ethnic minority groups and women whenever this is relevant?

Aspects for young children:

- Are there three-dimensional aspects to the display to encourage
- tactile experiences for pupils?
- Is the work of every pupil represented in the display?
- Is there a mixture of commercial posters and children's work?
- Does the display extend language development for pupils?

Aspects for secondary pupils:

- Does the display encourage cognitive development in scientific thinking?
- Is the work of pupils of all years celebrated?
- Are stimulating questions asked in connection with displays?

The purpose of display is to enhance the learning environment for pupils as well as to value their achievement. A well-planned display, which encourages pupil involvement in decisions about its construction can be a source of constant challenge. However, although there are examples of good practice in different phases of education, we need to consider whether these examples readily transfer between phases. Our criteria above indicate that there are issues of progression between the phases. Progression might be couched in terms of valuing, stimulating and imaging.

Valuing

In primary schools, it is important that the work of pupils is valued by being displayed, 'warts and all'. As pupils get older, their work needs to be redrafted for presentation, so that the 'warts' are ironed out. Sometimes this is done by presenting original handwritten work word-processed, so that redrafting has involved a further skill. For post 16 pupils, the whole issue of valuing may well have disappeared and what replaces it is one of communicating information and work being done to other students.

Stimulating

This is the opportunity to make display work for the teacher. Commercial posters, for example might have questions attached to make pupils think about the content, for example, labels attached to photographs of local pollution might read 'what conditions created this situation? What action would be needed to change it?' With young children the opportunity to raise questions about their environment — pictures of clouds for example asking 'what weather would we expect if we saw these in the sky?' or drawings of an electrical circuit with a caption; 'will it work? why?'. As children get older, the need is to have more searching questions that relate — for example — to variables operating in a particular situation, or pose problems to be solved. All this can be achieved through a caption which asks the relevant question.

Imaging

The exposure which children receive to images of people and science in society can be balanced through the careful use of display which helps children to question the 'taken-for-granted' aspects of television commercials and advertisements in general. Attaching the question 'What impression does this give us about the role of women?' to a poster of male scientists in white coats can begin to make children more aware of the underlying issues of equality of opportunity. Again, as young people grow up, the questions can become more challenging and work at the root of the image of science itself. Clever and careful use of questions on scientific experimentation or evidence can begin to introduce many of the contents required in science by raising awareness of the philosophies underlying science and the assumptions made about it. It is this trend towards the questioning of the philosophies of science which could form a valuable part of the stimulation seen in the displays of a sixth form science room.

Physical Surroundings

It is not uncommon for secondary science teachers to grumble about having to teach science in a room other than a specialist laboratory. In many ways it is a bonus to be grabbed and guarded. While the lab may have a wealth of materials at hand and promote the feel of doing 'real' science, fixed benches, taps and sinks, bottles and bunsens can also be unnecessary obstacles to learning. Similarly, we should be asking ourselves whether all science requires a laboratory, particularly in the light of the difficulties many schools are experiencing in planning to allocate 20 per cent time to all pupils for science at Key Stage 4. While we want children to feel at home in the lab, we want also to indicate that science happens in all parts of the world around us, not just with white-coated people in special rooms and conditions. In these days of modern science teaching, specialist accommodation is expensive and often underused. To be cost effective, every laboratory would need to have only science practical activities which demanded its specialist provisions in action every lesson. Activities such as discussion, role play, answering questions, reading, preparing reports would be non cost effective. The accommodation requirements of the next century may well be better met by a ratio of two laboratories to one classroom in science suites.

For our primary school colleagues, much of the reference to the physical environment of classrooms is very familiar stuff. They are used to different areas of the classroom, where the autonomy of children can be allowed more freedom, by allowing choice to be exercised. For secondary school readers, when work other than that of a practical nature is undertaken, a laboratory can be a real constraint. Fixed bench laboratories, for example make difficult conditions in which to conduct discussion groups. However, despite these obvious constraints, some flexibility is possible to facilitate learning. For us, the major aspects of any physical environment are that they should enable the development of autonomy in learners. We address this issue in more detail in Chapter 6, when examining the learning conditions which will assist in the management of assessment.

Cognition and Affect

In any environment the very act of learning is an emotional affair. The cognitive and the affective are not separate and distinct but are irrevocably intertwined. Learning brings with it a range of possible emotions from delight to fear, from satisfaction to frustration and despair. Our second strand of argument re-considers aspects of our constructivist philosophy. As constructivist theories of learning are translated into classroom practice (see, for example, Horscroft and Pope, 1985; Driver and Oldham, 1985;), the notion of a NTLE has grown in importance. The constructivism we discuss in Chapter 1 is one we have used in the activities throughout the book. In brief, it implies an arena in which youngsters are asked to consider the ideas and theories they hold for a particular topic, to explore these to some extent, to examine some of their consequences, to listen to and consider the ideas of others and to begin to re-shape their own ideas in order to take account of new factors. New additions or amendments to previous thinking might be brought about by either their teacher or peers.

When referring to conceptual change, some writers use the expressions 'cognitive conflict' and 'challenge', and it is not difficult to see that the process of cognitive change, within the robust community of a school classroom, could be very intimidating. In fact, the act of self-exposure of well held, or even tentative, personally constructed ideas and beliefs can be a daunting task and — in certain inhospitable circumstances — quite counter productive. Minstrell (1982), for example, suggests that for conceptual development to take place within a school situation there needs to be an engaging, free thinking, free speaking social context ... one in which students will put their thoughts up for consideration, free from fear of being chastised for being 'wrong'.

We would add that it might also need to be free of ridicule, supportive and empathetic of individual needs and emotions. No one, at any stage in life, can consider their own beliefs and theories coldly and dispassionately, particularly at the point of change. They are inevitably invested with feelings, not least the feelings of personal ownership.

In the remainder of this section we develop the notion of a NTLE in two ways. We expand upon the two threads we have outlined above, first, of conceptual change and second, the atmosphere within which it is conducted. In particular we focus on students' views of teacher behaviour. We consider data derived from interviews with school children concerning their perceptions of the ways in which teachers act in generating part of that atmosphere. Finally, we summarize the arguments and speculate on some of the implications of what has been said.

More about Conceptual Change

We have already outlined much of the recent debate within science education. Theories of conceptual change have been put forward by, for example, West and Pines (1985) and White (1988). These in turn have been digested and regurgitated through conferences (e.g., Secondary Science Curriculum Review, 1984), journals (e.g., *Physics Education*, 1985) review articles (Driver and Erickson, 1983; Gilbert and Watts, 1983), research programmes (Children's Learning in Science Project, 1990) and research bibliographies (Driver, Watts *et al.*, 1990). The gradual, passive, conceptual development of Piagetian theory (for example, Shayer and Adey, 1981) has given way to more overtly interventionist approaches. Part of that intervention is to shape and manipulate the circumstances that will facilitate individuals in the process of conceptual change.

The bulk of the research informing the debate has focused upon the gulf between learners' (like Cathy and Colin) own intuitive and personally constructed knowledge, and the formal, instructional, disciplined knowledge of schools. In our view the former is individualized meaning-making and is characterized by being tentative, personal and part of the learner's belief system. The latter is well developed, highly structured, of high status and can be characterized in terms of authority. As West and Pines (1985) describe it:

> It is 'correct'; it is what the book says; what the teacher says. It is approved by a whole bunch of other people who are usually older and more highly regarded than the student.

Creating a Climate for Communication

Conceptual change is commonly portrayed as taking place from personal intuitive knowledge to correct (scientific) knowledge. This sense of personally constructed meanings, varying between individuals and often at odds with the orthodoxy of school science, is at the heart of constructivism. Driver (1984) draws out a series of features of constructivism and argues it assumes:

> that learning outcomes depend not only on the learning environment but on the state of the learner, both on his or her conceptions and motivations. This implies that since learners may come to a learning task with different conceptions, they will learn different things from the same event.

She adds that constructivism also sees the learner as actively constructing his or her own meanings in any situation whether it is text, dialogue or physical experience, that construction of meaning is an active process of hypothesizing and hypothesis testing, and has the consequence that the learner is seen as being ultimately responsible for their own learning. Clearly, two youngsters can carry away two distinctively different perceptions of a teacher's actions.

A well articulated model for conceptual change is that developed by Strike and Posner (1985). They list four major conditions for a learner to undergo conceptual change:

- there must be dissatisfaction with existing conceptions;
- a new conception must be minimally understood, a person must be able to see how experience can be structured by a new conception;
- a new conception must appear initially plausible, to have the capacity to solve problems that provoked dissatisfaction in the old one;
- a new conception should suggest the possibility of being fruitful, of opening up new areas of thinking and explanation.

Swift (1984) has schematized this as follows:

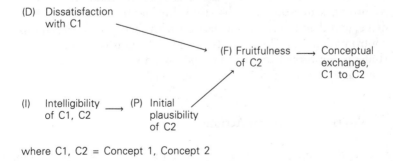

Figure 2.1: A model of conceptual change (Swift, 1984)

The process of change is not seen necessarily as being either linear or abrupt but may, for some students, be a gradual and piecemeal affair. Most classroom characterizations of this model of conceptual change see the teacher first assisting the learners to articulate and explore their own conceptions of some experience or phenomenon. The teacher then introduces anomalous features that are incapable of being easily interpreted by those conceptions and so induces dissatisfaction in the student. This kind of process is neatly summarized, for example, by Osborne and Freyberg (1985), and hinges upon the learner at some stage making explicit their own understandings either individually, in small groups or in a whole-class situation. They discuss four phases in the process: 'preliminary', 'focus', 'challenge' and 'application'. Focus, for the student, is the act of becoming familiar with classroom material concerning a particular concept; thinking and asking questions about the issue; describing what he/she knows about it, and presenting their own views to groups or discussion and display. Challenge is the consideration of individual's views by other pupils in the class, seeking merits and defects; the testing of the validity of these ideas by seeking evidence, and the comparison of these views with the orthodox scientific view. An example of this in action is that described by Task 4.6 where youngsters are asked to explain the behaviour of a gas in terms of the particulate theory of matter. One example of a model for such a teaching scheme is that by Driver and Oldham (1985). As teachers we see ourselves in the business of encouraging conceptual change. Unfortunately the conceptual change model as outlined above could have some undesirable side effects (Watts and Pope, 1985). For instance, Clark (1985) notes that

> The theory of conceptual change (as articulated by Posner, Strike *et al.*) holds that the state of readiness for conceptual change ought best to arise from the learner's own attempts to make sense of experience .... Yet a troublesome aspect of the way in which their work has been transformed into an instructional method is that the topics addressed arise from the wisdom of the curricularist, not the curiosity of the learner. The teacher is asked to rush the students to readiness by posing a question ... that probably never occurred to the students, and then induce dissatisfaction with their own explanations by confrontation.... The result is a kind of 'cognitive assault' in which students are forced to confront and abandon a part of self that has been, and is, serving them reasonably well.

Positive Non-verbal Actions

How then to mitigate the assault? What precisely might be a 'supportive atmosphere' of the sort we have advocated? More importantly, would youngsters themselves construe it as being a supportive one? Research into

Creating a Climate for Communication

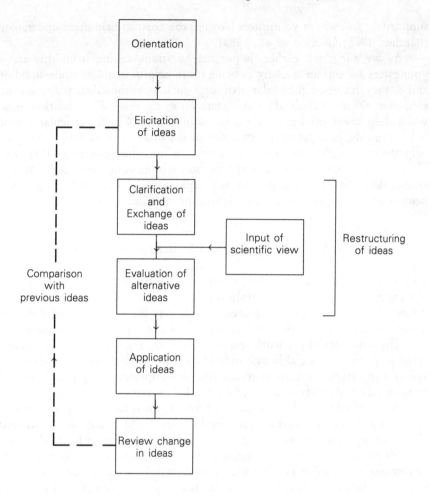

Figure 2.2: Constructivist teaching sequence (Driver and Oldham, 1985)

such aspects of support, through youngster's eyes, is rare. We can, however, build a picture of some aspects worth considering. Aspley's (1979) 'Kids don't learn from teachers they don't like', for instance, is pithily self explanatory. Or perhaps what youngsters require is teachers who — in Ausubel's (1964) words — can 'communicate a sense of excitement about their subject'; or Rosenshine's (1970) 'teaching behaviours which are desirable include enthusiasm, energy and surgency'. Rosenshine goes on to pinpoint some of the aspects of enthusiasm as being expressiveness, energy and mobility and identifies speech rate, frequency of gestures and smiling as indicators of these.

Other work in this area suggests that teachers who are perceived as being warm, empathic and enthusiastic human beings with a keen interest in youngsters and their ideas, are also seen as those teachers who are most supportive, and whom youngsters like and can trust to help them understand, (Bentley, 1985; Charkin *et al.*, 1983).

As we suggested earlier, important questions arising from this are: if youngsters see certain teachers as being more helpful in aiding understanding, and if they fix upon particular attributes such as enthusiasm, trust, warmth etc., then what precisely do youngsters mean by this? What do they mean when they say a teacher is warm or enthusiastic? Which particular teacher behaviours do youngsters use in order to construe them as people who will help them to understand new ideas, with whom they can share their own ideas without fear of ridicule, and who will join in youngsters' learning with enthusiasm? In the next section we explore some data indicating which behaviours youngsters find most significant in aiding understanding.

Youngsters' Constructions of a Supportive Atmosphere

Part of our work (Bentley, 1985) has been to interview youngsters to explore the meanings young people place on various kinds of teacher non-verbal behaviour. The interviews explored the reasons for selection and drew out aspects of teacher behaviour which helped to convey particular meanings.

The outcomes of the work relevant to our arguments here can be summarized as follows; first, although individuals' non-verbal behaviour can range across a spectrum from idiosyncratic to commonplace, young people are able to recognize accurately and describe intended meanings. As one might expect, they are shrewd and adroit judges of posture, expression and para-linguistic cues. Second, they use this information both implicitly and explicitly to guide their own activities, to gauge their entry, for example, into a teacher's awareness, or to assess teachers' evaluative feedback on their own actions. In many circumstances youngsters will judge a teacher to be calm, efficient, empathetic, sympathetic, warm, trustworthy, interested, enthusiastic, friendly, likeable and any combination or polar opposite of these.

We suggested earlier that youngsters see warm, empathetic and enthusiastic teachers as being those for whom they would most want to work. Some behaviours can be isolated which indicate warmth or enthusiasm, and are characteristic of teachers who are likeable and like them. It comes as no surprise to learn that youngsters see teachers who share jokes, smile, and lean forward as being friendly and likeable people. What is surprising is the degree to which youngsters can discriminate. For example, Sally, made precise judgments about particular expressions from small facial details — the expressions in the eyes, the set of the mouth:

> ... I can tell she likes the person she's talking to 'cos of the way her mouth kind of ... you know lifts up at the end. It's sort of like a joke but not a joke. Sort of a smile really, but not so's you'd really call it that.

This is quite a careful distinction between what is a joke and not quite a joke, through interpreting the set of the mouth. Presumably, faced with such a teacher, and with other cues, Sally can predict particular attitudes — and then control her own behaviour accordingly.

This fine attention to detail is not reserved just for the more obvious aspects of classroom life such as displaying liking. Paul and Sanjiv were each (separately) shown slides entitled 'Talking to a pupil who doesn't understand' and picked out the following aspects of the teachers' expression:

Paul: You know he (the teacher) is being sympathetic to the kid ... you know the one that doesn't understand ... and he's not putting him down, you know like telling him he's an idiot, 'cos his expression is understanding.

Interviewer: Which bits of his expression tell you that it's understanding and not putting someone down?

Paul: Well, if I could hear his voice I'd know really if he was putting him down. You know, it'd be sort of sarcastic and sharp. But even without it you can tell. It's his eyes. They look warm, they're sort of half smiling, and his mouth's relaxed and up at the corners. His face isn't tense like it would be if he was irritated.

Sanjiv: Look at how she's sitting. It's obvious in't it? She's sort of relaxed and friendly leaning forward as if she's interested in you and what you've got to say. You can tell she's not uptight. You'd want to talk to her normally wouldn't you ... not like a teacher you know. Sort of ... well a bit like my mum really. When you go in after school, and she's ready for a chat about the day. She sits relaxed like that.

The pictures were not of teachers they knew, so both youngsters used clues — accurately — to ascribe particular attitudes. Words like 'open' and 'trust' mean that, on the whole, youngsters believe teachers like and trust them:

Corrine: Yes, I'd think she liked me. She's lit up sort of ... in her face. You can see ... like as if she's going to giggle. Sort of ... warm ... like black faces are.... I expect she'd believe me ... you know not like that other slide. You could tell he didn't believe what he was hearing. He looked so tight! But she'd believe you. She's got a kind of honest face.

Interviewer: What is it about her face that's honest?

Corrine: I don't know ... she just has ... you always think

that when someone looks you in the eye don't you?
It's sort of open and friendly.

'Warm, friendly, open' teachers, then, are those who are active, lively, smile, display tolerance and are not tensely preoccupied with social control. They engage in frequent eye contact, listen carefully to youngsters' viewpoints, their actions indicate they are thinking these over carefully, and rarely use raised or angry tones. And are:

> Gary: enthusiastic. They care about what you think.
> Interviewer: How do you know when someone's enthusiastic?
> Gary: Well ... I suppose ... they're quick ... you know, they have little excited body movements.... Their eyes light up when they talk to you. Their speech is fast. They smile a lot, and joke. They seem ... well sort of fast, somehow ... their hands move quickly, they use their body a lot when they're talking ... they're not well ... still.
> Interviewer: And that makes you trust them?
> Gary: In some ways I suppose so. Well, trust is because ... well two things ... they make mistakes, you know enthusiastic people often do ... they're in such a hurry ... but when they do, they apologise. Listen carefully when you explain where they've gone wrong....
> Interviewer: How do you know they are listening carefully?
> Gary: They look at you, and they nod ... sometimes they sort of ... shake their head ... not saying no, but sort of considering ... and they ask you questions ... not angrily and aggressively.
> Interviewer: And that's why you trust them?
> Gary: Not just that. They show they trust you. They listen to your criticism. They're warm and understanding when you get it wrong.

Trust, then, is a process of sharing which requires mutual listening and a preparedness to act on what is heard. Pupils are clearly able to distinguish too between non-verbal and paralinguistic features of communication. In his vein Gary continues:

> Gary: Well ... it's ... sometimes the words can be different ... you know seeming to put you down ... but you know they're not really doing that ... the teacher actually likes you.
> Interviewer: Can you give me an example?

Gary: Well ... It's sort of ... well, when my computer program I wrote didn't work, and I said to Mr----- that I was having trouble, he said 'Not again, You're really being a twit with this one aren't you? Haven't you cracked it yet? And here was me thinking you were clever and going to get your O levels!' Well ... I mean if you just read that or I just told you it, like now, you might think he was being really unpleasant. I mean I can think of some that'd say that and mean it!

So....

Arguably most classroom acts of communication are geared to generating an effective working (learning) environment, reaching apotheosis in a warm and trusting atmosphere. Normal school settings are where students are encouraged (required) to make explicit their existing ideas in order that ensuing explorations will find them wanting. As Strike and Posner (1985) say:

> Dissatisfaction with the existing conception decreases its status, while exploring the fruitfulness of an alternative conception increases the alternative's status.... Therefore, competition between conceptions results in a process of accommodation characterized by temporary advances, frequent retreats and periods of indecision.

For us, this notion of cognitive change relies upon the development of a supportive classroom climate. The exposure of personally held ideas and beliefs, of personal 'retreats' and 'indecision' requires an atmosphere of warmth and trust.

From the discussion and examples we can summarize a series of points:

- youngsters are highly attentive and sophisticated interpreters of teacher behaviour;
- within a supportive atmosphere they commonly look for high levels of trust, warmth and enthusiasm;
- they know they have found such attitudes when teachers engage in frequent eye contact, are alert, with quick body movements, listen carefully to youngsters' criticisms, and act on them;
- such teachers laugh, with others and at themselves, stand close, touch from time to time, and use quick, bright voice tones that convey warmth and above all, respect for their co-learners.

These skills may — or may not — come naturally to individual teachers. In any event, successful teachers work hard at developing classroom 'personalities' and know when they can slip from one type of persona to another. While their own underlying personality is abundantly clear to all, their 'act' communicates its own layer and level of meaning to the class.

Similarly, the skills of decoding these behaviours and levels of meaning may or may not come naturally to young people. While some can interpret — and use — their own body language and presentation to great effect, we nevertheless believe there is much teaching to do. In other chapters we look at some issues of role play, but here we want to develop activities concerned with direct skill teaching. In particular, the need to teach children ways of talking and listening.

Direct Skill Teaching 1: Talking

At once, talking is the most simple and yet complex process open to us. Pupils' talk in school has received close attention for twenty years or more: starting from, say, the seminal work of Barnes, Britton and Rosen (1969), Flanders (1970) and Wilkinson (1971). Because of the enormous amount of work accomplished in the area there is a temptation to think that talk — both the 'insider' talk of classroom peer groups and the 'outsider' talk used to make conversation 'fit for purpose' — has reached a point of saturation with little more to offer. However, the introduction of the National Curriculum has brought a re-awakening of the debate. As Ogborn (1991) points out, the publication of the Science Document raises some questions over the assumptions scientists make about the prior skills and competencies children have at various stages, and whose role it is to teach these. There can be no doubt in our mind that scientists are vitally involved in the direct teaching of communication skills, let alone making use of skills developed by other teachers in the school. To assist the teacher of science, Ogborn would add the following questions to those in the NCC guidelines:

- Is talk (in this lesson) the appropriate means of learning (compared with listening, reading and writing)?
- Is it the starting point, from which activities will develop?
- Is it to be used to demonstrate learning?
- Will it be whole class discussion (if so will it really be the whole class, or just one or two)?
- Will girls and boys speak or just boys?
- Will it be pupil talk or teacher talk?
- How much time will they spend in pairs or small groups?
- How will presentations and feedback be organized?

Let us explore an example. A group of 15-year-olds (Key Stage 4) are discussing animal experimentation. The teacher has been aware of some divided opinion in the class, possibly prompted by a playground incident when boys in the class teased and mistreated a cat. She wishes to explore some of the pupils' attitudes about respect for other life forms, and chooses to begin the discussion by taking some facts and figures from published material, principally 'Victims of Science: the use of animals in research' (Ryder, 1975). A telling passage is:

> Research on animals used to be almost exclusively a medical affair, but nowadays an increasingly large number of non-medical fields also use animals; some of the largest and most obvious medical fields combined together have, over the years, accounted for only one third of all British experiments.
> Commercial examples of non-medical experiments are the oral toxicity testing of weedkillers, packaging materials, cosmetics and toiletries, food-dyes, flavouring additives, detergents, floor polish and anti-freeze liquids; in such experiments these substances are force fed to animals to kill them.

Pupils were organized in mixed gender groups and mixed, too, (as far as she could tell) in their opinions on issues of 'animal rights'. She changed from group discussion to whole-class discussion within the single lesson period available to her. Each group had a number of questions to which they were required to give answers — agreed answers where possible, opposing points of view where not. However, she had a further 'teaching point' to be developed: she wanted youngsters to become aware of their own feelings and pay attention to those of others in the group. Having set the scene for the discussion, she asked pupils to read a selection of short articles and passages like the one above, and then spent five minutes outlining her *'today's* rules for the discussion'. These were:

1 Own up to your feelings. If you feel angry or disgusted, they are your feelings, don't cover them up. Say 'I feel angry'.
2 Don't blame others for your feelings. Don't say 'You make me feel disgusted, when you say that'. Say something like 'When you say that, I feel disgusted'.
3 Don't speak for others in the group, particularly about their feelings. Instead of 'We are all fed up with that argument', say 'I feel bored with that, how do you feel?', or 'I'd like to move on'.
4 Avoid judging others. In place of 'You are wrong', use 'I disagree with you'.

The class maintained the rules themselves and were later able to note points where individuals found it difficult to keep to them. They approved of the rules even while finding them difficult to apply, particularly the notion of not blaming or dismissing another speaker for engendering strong personal feelings.

Other group discussions can be geared to other learning points, for example, turn taking; time sharing (say, only one minute's talk per person); word counts (provide a description in no more than fifty words); two points 'for' and three 'against'; even to the effect of deciding on what *is* a good discussion in the first place (Watts, 1989).

Solomon (1989) reports the work of a research student (Wallace, 1986) who describes pupil-pupil talk at Key Stage 3 during the process of practical work in the laboratory. She categorizes the talk in six ways:

1 Negotiating doing (e.g., arranging the collection of apparatus and turn taking in the experiment)
2 Removing tension (e.g., when disappointments or near quarrels have occurred)
3 Giving help and tutoring
4 Non-task talk (e.g., greeting and 'stroking' as they settle into pairs and trios)
5 Negotiating knowledge (e.g., agreeing or disagreeing about what colours, measurements or tastes they perceive)
6 Constructing meaning.

Wallace makes the point that this kind of talk is necessary for the smooth running of the group, whether this entails organizing the physical space within which the activity occurs or arranging the social and cognitive milieu surrounding the task.

Similarly, Playfoot (1990) highlights the impact of pupil interaction in small group problem-solving. Both the teacher's input concerning the problem and the cognitive and technological demands of the problem, compete with peer interaction as variables in the success of the group in deriving a plausible solution.

Although we give only one example here of direct skill teaching as far as talking is concerned, there are several other examples throughout the chapters that follow of pupils acting as group leaders and having to instruct a group in what is to happen, and of reporting back by members of a small group to the whole class. One other interesting example seen recently in one very good English lesson came when an adult had been invited to tell the class of her experiences during the war in an occupied country. This fitted in with the work which they had been doing on a particular book. The adult addressed the whole group, then the class broke up into small groups and decided upon one question which they could ask of her to help them understand better the characters or plot in the book they were reading. Many pupils focused upon political issues of today with regard to the occupied country, because they were topical and of interest, rather than following the brief given about their book. After the visitor had gone, the teacher helped the class to see how the question might have been phrased to find out both topical information and focus attention on their book. He reminded pupils of the reason for having a visitor, that they could collect a type of 'evidence' about feelings etc., which was more immediate than that found in the book. He helped them to frame more questions for the visitor in ways that would help her understand what they wanted and be able to answer them constructively. She was then invited back to discuss the questions in a further lesson.

Although this example comes from a different area, it reminds us of the preparation needed if pupils are to find out precisely what they need in an interview situation. This activity might have been followed up in a science lesson by focusing on a piece of survey work and asking pupils to design questions to elicit certain information. The surveys could have been tried out and the results and experiences analyzed, perhaps with some of the participants of the survey present.

Direct Skill Teaching 2: Listening

Research (reported, for example, by Wilkinson *et al.*, 1974) suggests that we listen very badly, which probably explains why children are constantly harangued to 'listen closely', 'pay attention', 'watch my lips — I will say this only once'.

Students listening to lectures have been found to comprehend half, or less than half, the basic matter of the lecture. Of course, that may say as much about the quality of the lecture as it does about student powers of listening, but 'passive' listening is also to blame — the kind of listening that requires little more than human presence. In contrast, we might say that *active* listening requires immediate action to be taken on the information gleaned.

While some aspects of teaching and learning have changed for the better, some of the lecturer-listener ways of working are still around. As Abercrombie and Terry (1978) have pointed out, it is not uncommon for teachers and student teachers to arrive for seminar, tutorial and discussion sessions prepared to listen passively but not to contribute actively to the theme in question. There are numerous occasions when active listening is essential for a comprehensive appreciation of what is happening. Listening, of course, is as vital to a conversation as talk not least, as Beattie (1983) points out, for regulating the conversation and turn taking. One has to listen to know when to enter without simply talking over or shouting down the speaker. The listener is important to the speaker: their gaze and body language is a constant stream of comment on what is taking place. One only has to recall how different conversation is by telephone to appreciate the point.

A lesson watched recently made use of a 'wordless demonstration': the class (Key Stage 2) each had a worksheet upon which were the steps and processes of an experiment to observe the effects of temperature on the rate and size of crystal growth. It is a straightforward procedure where some few grams of salol (phenyl-1, 2-hydroxybenzoate) are first melted (in a test tube in warm water) and drops are then placed consecutively on a warm microscope slide and then a cold one. The crystals can be watched as they grow through a hand lens or low-power microscope. However, on the worksheet the steps were placed in the wrong order and the class were asked to watch as the teacher undertook the demonstration. The teacher said nothing throughout, but one pupil — without the benefit of the worksheet — was asked to describe out loud what the teacher was doing so that the others — quiet and listening — could put the steps on the worksheet in the right order.

The longer we are required to listen passively, the more difficult it becomes: twenty minutes is one thing, two hours and twenty minutes quite something else. It is very unrealistic to expect learners to maintain maximum concentration for prolonged periods of time; the school day needs to be interspersed with a variety of listening tasks. Sutton *et al.* (1981) distinguish four levels of listening:

1 Casual listening. At this lowest level we are aware that certain things are happening: hearing background music is a good example;
2 Conversational listening. This is the kind of informal chat, small-talk listening that is aware but requires little effort;
3 Appreciative listening. This means giving something one's undivided attention: listening to directions, a passage of music, a bird song, a play etc.;
4 Critical listening. This highest level implies intense concentration: listening for a mechanical fault in an engine, a clarinet teacher listening to a pupil play, grappling with news on a particularly powerful occasion, or the meanings of a person in a language which is not your first.

These are really gradations in distinction between hearing and listening, something which can be brought home to youngsters as they work in school. For example, ask very young children to observe closely (look and listen to) an earthworm as it moves across a piece of paper and, in describing their observations, to say why it happens. Hearing the noises made is distinct from listening to what is happening — the latter requires some understanding of the mechanisms involved.

This kind of experience can be a prelude to investigations of the senses of hearing and listening. A well tested idea is described in 'How Scientists Work' (Lythe, 1986) for Key Stage 2 or 3. A large clock, one or two metres radius) is drawn in chalk on the floor and the hour numbers marked in. A volunteer is blindfolded and sits at the centre facing twelve o'clock. A partner moves quietly round the outside of the clock and periodically stops to tap two coins together. The blindfolded child must point to where she/he thinks the noise is coming from and the correct and incorrect positions are noted. Partners are changed and sets of results are pooled. The youngsters are asked to say what this tells us about the ears' use in direction finding. The rest of the room must be suitably quiet, listening, while the exercise takes place.

The Further Education Unit (FEU, 1988) lists a number of attributes for effective listening at least, say, at the 'appreciative' and 'critical' level, such that effective listeners:

1. display appropriate non-verbal behaviour, adopt a positive posture, maintain eye contact, nod and smile when appropriate, and focus attention on the speaker;
2. accept ideas and feelings ('That's an interesting idea, can you say more about that?'), and can empathize with the speaker ('So when that happened you felt angry');
3. can probe in a helpful way ('What led you to feel that way?');
4. paraphrase at appropriate times to check understanding, and summarize the progress of the conversation from time to time;
5. can add to the comments being made, can widen the range of ideas by suggesting a number of alternatives.

A favourite game for any age and stage is to organize pupils into groups of three: one 'the talker', one 'the listener' and the other 'the observer'. The talker and listener sit back-to-back and are told to stay that way for the duration (5 minutes) of the game. The talker takes about two or three minutes to tell the listener some humourous story, anecdote or joke; the observer must note down the non-verbal behaviour of both. What commonly happens is that, as the story gets underway, both speaker and listener begin to turn their heads towards each other. It is very difficult to tell a funny tale without feedback from an audience and it is also difficult to appreciate humour without access to the facial expressions, gestures and movements of the humourist. The exercise highlights the need for a listener's 'positive posture' and attention. A further step in the game is to ask the observers to share their notes with the whole class and for the teacher to write the essential points on the board. Re-arrange the groups into different trios, this time to work face-to-face. Now, for five minutes, the talker tells the listener about a particular time she/he had a difficult problem when someone close or in authority would not hear what they had to say. The observer should use the notes on the board as a checklist and watch how the listener listens. When time is called the observer tells the listener what was seen and noted. Then the talker comments on the listener and, finally, the listener gets right-of-reply to say how easy or difficult it was to listen.

Another example of teaching for listening is 'Chinese Whispers', a well-known game where a message (traditionally 'send reinforcements the enemy has advanced') is whispered from person to person round the room, the last person having to say aloud the message that eventually arrives. It is valuable to repeat the game, this time asking the pupils to note down what each hears before relaying the message onwards. This is the systematic way a scientist might adopt — not just to marvel at the phenomenon but to try to track the process as it happens. There are many more activities, for instance:

ACTIVITY 2.1

When you are using a video or a film, make sure the class has a list of questions that can be answered by listening to the sound track. Commonly educational videos come with a pamphlet giving a fairly close approximation of the action and the commentary. Ask the class which questions on the sheet helped most.
If the narrator or commentator does not appear, ask the class to decide what sort of person she/he is from the sound of the voice.

ACTIVITY 2.2

Occasionally use entirely inappropriate wrong or nonsense words in a description or account — did they spot the 'deliberate mistake'? The class will only be really alert to this kind of exercise if you have already undertaken some 'listening activities' with them. Be careful not to make the word too realistic for some who may want to adopt it as scientific and not appreciate the joke.

ACTIVITY 2.3

Audio cloze — read the instructions for an experiment onto tape and omit words as you go along. This can be (say) every fifth word regardless, or significant words as you choose. You can either prepare a worksheet for them to complete or expect them to make notes as the tape progresses. Make sure to check their interpretations before safe practical work can begin.

ACTIVITY 2.4

This is called 'Sentencing'. On a small piece of paper all pupils write a sentence relevant to the topic in hand. These are folded and put in a 'hat' somewhere up front of the class. Now form into groups of three, four,

> five — the number over three is not critical. Two 'contestants' each take a sentence from the 'hat'. The object of the game is for the contestants to discuss the topic in hand and slip in their chosen sentence unbeknown to anyone else. When you call 'time' (two or three minutes or so) the observers have to guess which was the sentence each contestant had drawn.

Science and Bilingual Pupils

Communication skills in science, of the type we have outlined above, take on a particular poignancy when the needs of bilingual learners are considered. One of those needs might well be to acquire a level of fluency before coming to terms with science in general and the National Curriculum in particular. However, the Hounslow Languages support services team are firm in the belief that this is not a way forward:

> exemption, even for a limited period, from a science course for a pupil with little or no English is something we do not recommend.... The science classroom is one of the best environments ... for promoting good language development (Roach *et al.*, 1990)

Ellis (1984) outlined several factors as being important for developing language for second language learners. The CLSS team regard these as being particularly pertinent in science:

Facilitating the Need to Communicate

Bilingual learners need to communicate with the teacher and other learners in order to reach understanding and be certain of the tasks and their role in them. Science activities, if well structured and organized so that collaboration — for example in groups — is part of the activities, can aid this need to communicate.

Linking Words Closely to Activities — The 'Here and Now' Principle

Science, because of its close relationship with practical activities provides an excellent vehicle for matching words with objects which bilingual learners can observe directly, or events, which they can watch and these help when learners come to try and express the more difficult explanatory ideas of science.

Communicating in School Science

A Directive-rich Environment

Science provides instruction for young people. One of the statements of attainment in National Curriculum Science is to be able to follow instructions. Because of its close linking of words with action, science is an excellent medium for providing the words and action environment that bilingual learners can gain so much from.

Making the Most of the Opportunities Science Provides

In providing the aspects outlined above, science can be an excellent learning medium for second language learners. However, adaptation of the environment to take account of bilingual learners is necessary if communication skills are to be enhanced.

When planning experiences for bilingual pupils that are designed to help them in developing their skills of communication, three phases of lesson construction might be exploited.

Instructional phase: during these activities, teacher or pupils might be talking and planning and listening to one another. Structuring this with particular listening activities can help bi-lingual learners.

Planning and carrying out investigations: during this phase, structured activities which direct pupils' attention towards the processes of science — hypothesizing, predicting, testing will help the understanding of these terms and recognition of them in future. It will have the added advantage of ensuring that children's talk in groups is not just of the substance of what they are doing, but includes interpreting observations, relating results to theory and future planning and testing.

Consolidating: writing up in a way which encourages scientific understanding and the use of both scientific vocabulary and register.

There is an example of an activity which might fit into each phase, drawn from the work of the Community Languages Support Services Unit in Hounslow (Roach *et al.*, 1990).

Listening Activity

This activity has been chosen to illustrate how any piece of experimentation which is introduced by the teacher prior to pupils being involved can use verbal instructions to help develop the listening skills of bilingual learners.

Creating a Climate for Communication

The teacher's instructions are accompanied by a set of visual instruction cards, which the pupils are required to order in sequence as they listen. This encourages active listening and ensures that pupils do what is required of them. It will of course only work with those activities where the teacher is giving precise instructions, and not for more open-ended activities. However, bilingual pupils can be encouraged to listen actively in these too, by drawing the instructions as they are decided upon and then working with a partner later to match visual instructions to the partner's written ones. This type of pairing also has the advantage of helping the bilingual pupil to see how to write simple clear instructions of an activity they have designed and carried out. Figure 2.3 below shows the instructions and picture card that a teacher might use for dissolving copper sulphate.

Talking

The example given here (Figure 2.4) is a simple exercise, which will assist bilingual and other learners to focus their attention on the processes of science as they work during an experiment. As it focuses around a group outcome rather than an individual one, it encourages discussion between group members and forms a final 'blueprint' for writing up the results by individuals, since each observation and prediction can be linked to a simple sentence which bi-lingual pupils can adapt to different uses. For example, the work sheet could end with these simple sentences.

We predicted when the switch was off, bulb --- would light.
We predicted when the switch was off, bulb --- would not light.

We observed when the switch was off, bulb --- would light.
We observed when the switch was off, bulb --- would not light.

Further sentences could be developed for each stage of the experimentation by the pupil, drawing on the format of the two provided by the teacher. This helps bilingual pupils to experience the sentence structure used in scientific reporting of results and practise it.

Writing

There are two examples here. One is used to help pupils structure what they have done by referring to instructions and then helping them change the tense of their work into the past, so that the account takes on a particular format. The other — concerning a similar area of work, is to help them write more complex sentences by using linking words of various types to show rela-

Communicating in School Science

Preparation of Copper Sulphate

The picture cards are cut up and distributed, one set to each group or individual, according to the aims of the teacher.
The teacher reads out the instructions step by step, repeating each one.
The pupils, in groups or individually, sequence the picture cards as they listen.
Finally the teacher rereads the instructions quickly and gives out written instructions.
After checking, groups are allowed to carry out the practical.

Instructions to be read by teacher:-

1) Heat about 25cm^3 of sulphuric acid in a small beaker.
2) Add two spatulas of copper oxide.
3) Stir the mixture.
4) Decide if all the copper oxide dissolves.
5) Add two more spatulas if all the powder dissolves.
6) Stop adding powder when some copper oxide is at the bottom.
7) Filter the mixture into an evaporating dish.
8) Heat the solution in the evaporating dish until the volume is halved.
9) Cover the evaporating dish with paper.
10) Leave it until next week.

The picture cards.

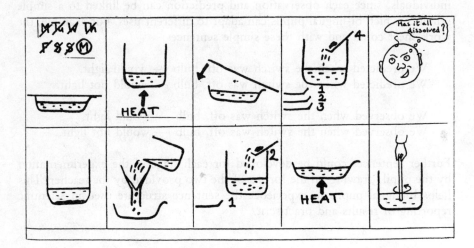

Figure 2.3: *Workcard for an experiment with copper sulphate*

Parallel Circuits

Circuit	Switch	Prediction	Observation	Explanation
1	off			
1	on			
2	off			
2	on			
3	off			
3	on			
4	off			
4	on			

Talking exercise — Parallel circuits
This exercise leads on from the listening exercise, 'parallel circuits'.
Each group of students is given a copy of the results and prediction worksheet.
Before setting up the circuit, they are asked to make a group prediction for what will happen to the bulb(s) when the switch is turned on. After writing this down they set up the circuit and record their observations (if desired, an extra column asking for an explanation could be added).
They then repeat the process for each circuit in turn. It is expected that their predictions will improve as they take into account their successive observations.

Figure 2.4: Workcard to assist bilingual students

Investigating a mixture of starch and diastase.

Pupils' instructions (given by teacher orally, step by step)

While you are waiting for changes to take place in your experiment you can begin to write it up.

i) Working with a partner, arrange the instruction cards so they are in the right order.

ii) Write down on rough paper all the verbs which you will have to change when you write in past tense saying what you have already done.

iii) On the rough paper write past tense verb next to the verb from the instructions.

e.g.
Tie — tied.

iv) Start to write your report in past tense, by changing the instruction cards and adding the time connectors in the box where you think best.

v) In your report, include anything you did to improve the experiment that is not in the instructions.

Instruction cards.

Tie a piece of cotton round one end of both pieces of Visking tubing.
Tie both bags at the other end.
Add water to one bag until it is three-quarters full.
Put both Visking tubing bags into boiling tubes containing water.
Stand the boiling tubes in a beaker for about forty minutes.
Add diastase to the other bag until it is three-quarters full.
Collect two pieces of Visking tubing, two boiling tubes, a beaker and some cotton thread.
Half-fill both Visking tubing bags with starch suspension.

Time connectors

First ; Then ; Next ; After that ; Finally.

Figure 2.5: Workcard for bilingual students to assist with writing-up experiments

Creating a Climate for Communication

> Digestion
>
> We use these words to show how two ideas are related.
>
> | and; because; but; if; in order to; so that; even if. |
>
> Use them to complete these sentences.
> a) The small intestine absorbs food easily _____ it has thin walls.
> b) The gullet has muscles _____ food moves quickly to the stomach.
> c) Soluble food goes into the blood _____ insoluble food is not absorbed.
> d) _____ you grind food into a powder it won't dissolve.
> e) Food passes out of the anus _____ it is not digested.
> f) Digestion takes place in the stomach _____ small intestine.
> g) Muscles mix food and digestive juices _____ speed digestion.
> h) Breakdown of food happens _____ digestive juices are added.
> i) Food goes through the gullet _____ is not digested in it.

Figure 2.6: Workcard to assist bilingual students with written work in Science

tionships between ideas, such as cause and effect etc., which are commonly occurring aspects of science.

In general, assisting bilingual pupils to understand listening, talking and writing skills in science is not dissimilar from helping pupils whose first language is English. There are points about the register and vocabulary of science which need attention, and opportunities for practice, just as there are aspects of scientific understanding which need attention. We believe that such activities can be greatly enhanced by helping pupils to draw on the expertise of others in the room — adult or peers. Setting up a framework for collaborative group work is an important feature. We focus on this in our next chapter.

Chapter 3

Communication and Groupwork

Introduction

> To be human is synonymous with being in communication and relationship with other people, which demands of us a range of social skills. (Button, 1983, p. 1)

In the previous chapter we looked at aspects of the learning environment for the whole class, and the teachers' role in establishing what the norms might be for the class. However, not all teaching is whole-class teaching, there are many other arrangements for learning.

This chapter explores groupwork as a tool in enhancing communication in science lessons. It focuses particularly on the 'group' as the mechanism for ensuring that ideas are explored, understood and acted upon. The nature of cooperative and collaborative learning in groups is under scrutiny.

Our main emphasis in Chapters 1 and 2 was 'communication for concept development'. In this chapter the emphasis is 'communication for concept development through small groupwork'. We noted Vygotsky's work which tethered concepts and language: the concepts we form and change throughout life would be of little use if we could not communicate them within our social environment. The Greek derivation of communication ('to make common') indicates the intention to share meanings. Our vehicle in this chapter is the 'small group' and we take pains to define what we mean and how we arrive at particular combinations of individuals to form groups. The latter part of the chapter explores the skills implied by the document 'National Curriculum: Science'. The chapter contains more activities, where we try to give examples of the organization of effective groups, and pursue our argument for four main criteria for establishing and maintaining groups.

Groups in Science

There is no doubt that the majority of science teachers would regard small group work as an everyday experience. There is no doubt, too, that many science teachers have long encouraged pupils to work in a cooperative way towards the completion of a particular experiment and task. Here we try to extend those horizons, by looking (sometimes outside science) for new ways of using groups as a major element in teaching and learning techniques.

Almost every teacher in initial training has heard the phrase 'peer group pressure'. It implies that individual youngsters somehow come under the influence of others in their class or group — usually in ways that are less than helpful to the general enterprise. Peer group pressure has a negative connotation, something to be aware of and, if at all possible, avoided. Usually it is seen to subvert constructive work, it results in hard working youngsters being 'led astray' by their peers (into smoking, drugs and bad ways generally). Yet, despite the power peer groups are supposed to wield, few training techniques offer helpful advice on how to harness that power and use it to further learning. We cannot re-examine all issues of peer group pressure here, though we are convinced that groups are a powerful force in any classroom. Our intention is to step back, examine some of the processes which make groups work and then explore how we might use these processes to ensure learning science can be enhanced.

Why Groupwork?

Since this is a book about communication in science, one might well query the relationship between groupwork and communication. After all, SATs (Standard Assessment Tasks) are designed to assess the attainment of pupils as individuals, not the achievements of groups. That is, group communication skills still need to be taught, and children need to learn to work together even if this is not now to be formally assessed. For us, however, the issue is much more fundamental — if youngsters are to understand scientific ideas, they need first and foremost to share those ideas. There are few areas of science in which youngsters have no experience whatsoever, all of them come to science, at whatever level, with some understanding of their own. This understanding needs to be explored by talking, writing, experimenting, testing limits and making comparisons with the understanding of others.

Most science lessons involve pupils working in groups. Unfortunately, the rationale is rarely the need to capitalize upon the power of peer groups and still less to share learning through communication. Most usually, the need for group work in science is decided by the availability of resources. Few classes of twenty-eight pupils have twenty-eight identical sets of equipment for experimentation. Thus, pupils work in twos or threes because this saves on provision of materials. It is interesting to note changes in practice which have appeared with the advent of GCSE course work: pupils have been provided with equipment and expected to work as individuals, even if this has meant half the class 'doing the practical' while the remaining half do 'bookwork'. That is, scientific equipment — or the lack of it — is commonly the sole rationale for gathering pupils together: the development of scientific concepts *per se* has not been an issue for re-arranging working environments.

So, for example, rarely are the instructions for an experiment given in ways that lay down complementary roles for individuals in the group, nor is a sense of common purpose developed towards the completion of experimentation. We call this collective work — a collection of people working on an identical experiment. No part of this classroom organization draws on individual skills brought to the group task; seldom is there a sense of shared goals or purpose or deliberate framework which helps people learn from one another's strengths. Collaboration is commonly planned as collaboration in physical activity not conceptual enrichment.

To us, though, it is group work if it:

- involves two or more people working towards a common, and negotiated goal through a shared plan of action;
- designates, preferably by negotiation within the group, different roles for different individuals, which can be seen clearly to relate to the tasks in hand;
- allows every member of the group to bring their skills to bear on the common task;
- creates an atmosphere in which every individual has the opportunity to learn from every other individual in the group;
- encourages all members of the group to work to their strengths and be prepared to share and overcome their weaknesses in relation to the task;
- encourages a working environment of trust and corporate responsibility between group members.

Classwork and Groupwork

A group often starts out as being a collection of people with disparate experiences and interests who have come together for a particular purpose. In science education, there could be said to be two types of group:

1. the whole class. We can all remember some large groups — classes — with particular identities, quite unique and different from other class groups in the school — the 3c's and 4z's of the world, some exceedingly difficult to manage, others a joy to work with;
2. smaller subsets of the whole class created by the teacher, or the pupils

for particular purposes. The bulk of our discussion concerns the second of these, though we return to the first somewhat later.

The ORACLE Project (Galton *et al.*, 1980) based in Leicester from 1975–1980 characterized teachers by their propensity for groupwork in class:

Individual Monitors

These teachers use a high level of non-verbal interaction between themselves and the class, mostly using individualized instruction and then monitoring individual pupils' work by moving round the class or having youngsters queue at teacher's desk.

Class Inquirers

Here the pattern is of whole-class teaching, indicating very teacher-directed learning. There is commonly a high level of communication with the class and individuals, and of factual information and book-based problems.

Group Instructors

In these cases the teachers are seen to tightly structure learning activities, which are characterized by much verbal interchange and a mixture of information-giving and the use of open questioning techniques.

Style Changers

This final category is an amalgam of the other three and consists of three subsets:

a   'infrequent' changers who group pupils by either friendship or ability, and then make occasional changes to group structure when they need to alter tactics;
b   'rotating changers' who organize groups on a planned basis and systematically rotate members of groups, often organizing class sessions on a circus basis; and
c   'habitual changers' who make frequent unplanned changes between whole-group and small-group tasks.

We believe communication is an essential ingredient in the learning of scientific concepts. But holding hard to this position entails an immediate and powerful effect on the classroom arrangements we choose as we teach. In order to ensure that communication-for-concept-learning occurs we cannot leave it to chance, simply a 'hit and miss' affair. We must ensure every opportunity is taken to develop the communication of scientific ideas between pupils, between pupils and teachers, and between pupils and scientists and industrialists, or any other interested adult in the classroom. Here, then, using the Oracle terminology we are concerned with that style of teaching encompassed by Group Instructors and Rotating Changers. We are interested in closely structuring the sweep of classroom activities and regularly reshaping working groups in the tasks they have and the roles they play.

This means planning teaching with two aspects firmly in mind:

- the nature of the working arrangements
- the nature of the activities.

Both aspects — the working arrangement and the nature of the task — are essential, and closely wedded. Pupils frequently work together in classroom activities and do so in a variety of constellations; pairs, small groups, whole-class groups etc. 'Active tutorial' teaching (for instance the work by Baldwin and Smith, 1983) commonly uses pupils working in paired situations, often in order to develop the ideas between them. Many teachers are familiar with the techniques of 'pairs', 'fours' and 'sharing' (if only by having experienced them in INSET sessions) which are an integral feature of such work. However, simply organizing group experiences for pupils is not an end in itself.

And after all, is not a whole-class a working group? Yeomans (1987) has some interesting thoughts:

> I was sitting in a room with thirty others. Few people had spoken, ... I nearly hadn't bothered to come to the class. Several weekly sessions of this large group experience left me feeling isolated, powerless, insecure, uncooperative and hostile to those I blamed for my feelings.... Large group experience contrasted vividly with the small group experience later in the week ... the group of ten differed from the large group in size and configuration but not organization or structure. Both groups had a responsible person as a non-directing leader ... but the ten were sitting in a circle and able to see each other ... (this) led progressively to extensive talk, openness, supportiveness, warmth, cohesive interdependence and a sense of common purpose.

A powerful description indeed of the differences between being a member of a large group — of thirty — to one of a much smaller group, in which the 'rules' are similar, but the conditions create different circumstances: in the small group interactions can occur much more productively. In other words group size is an important factor in the experience of being in a group, and a whole-class group can have threatening connotations. This is particularly the case for young children, for some older ones too, especially in transition stages between phases of education. Pines (1975) puts this succinctly when he states that

> the individual feels lost and out of touch with aspects of the self and others; fears domination by the large group.... (S)he is unable to find a familiar or useful role and becomes deskilled, disorientated (p. 294)

Yeomans (op. cit.) gives a variety of examples of how, from the pupil's viewpoint, a class is a large group, in contrast, say, to a societal view of large groups (a football crowd, a pop concert). He quotes Turquet's (1975) definition of large groups as 'not being face to face, ... (where) the kind of personal interaction with other members and with the whole which is characteristic of the individual's experience in small groups is no longer possible' as an example of some pupils' difficulties with the whole-class as a feasible group size for work. There are a variety of circumstances when young people need to be managed in large class groups, an example might be watching a video to enhance a description or an explanation and our examples in the book suggest such methods fairly frequently. For many classes though (particularly in secondary schools) this is too often the predominant mode of operation.

What Does Creating Effective Groups Involve?

One could say that when any group of people work together over a period of time, then their identities and involvement develops 'naturally'. The assumption is that the act of grouping alone will ensure successful outcomes. Indeed, there is some research support for this. As long ago as 1897, Triplett involved youngsters in a fairly simple task — that of reeling in a fishing line. He found that an individual's speed at the task improved every time he or she worked together with someone else. There was no notion, at this point, of pairing the youngsters on grounds of skills in fishing or friendship, it was simply the presence and cooperation of another youngster which improved the performance. Further studies of such 'co-action' have taken place since 1897 over a variety of tasks and, on the whole, the findings have been similar, the mere presence of others cooperating in a task can act to increase the speed at which tasks are done. In other words, even just the act of putting two individuals together on the same task, with no thought of careful matching of task or people, can enhance learning. But we would stress that this is not what we are about. For us, the matching process is a very important part of group creation and is the first step to effective teamwork.

As we noted earlier, the nature of the task is important too and in some cases can actually interfere with the effects on performance. Allport's (1924) research showed that if pairs were asked to complete multiplication tasks, or look at a picture to spot a reversible perspective, both the speed and quantity of work increased. If, however, the same pairs were asked to produce good verbal arguments, their speed of performance was inhibited by the presence of the other person and the quality was inferior to that produced when working alone. One conclusion, then, might be to group youngsters for simple tasks but let them work alone for more complex ones. Clearly this is too simplistic: the nature of group processes needs a much more careful examination and we devote time to this later.

In other words, for us, a group is not just a collection of individuals defined by the frequency of their interactions or the fact that they are all working on a common task. It may, for example, be a group because the members share norms in some way and between whom there exist concrete, dynamic, inter-relations. Needless to say, there are many examples of groups that will not meet these criteria; not all committees, (or members of committees) for example share common norms and (silent) Trappist monks might be said to be a group with few opportunities for dynamic verbal communication.

Creating Groups

What is it then that all groups need to have in common to be effective — and what must we take account of in creating our groups in the first place — if we are to use groups to enhance learning? Gaskell and Sealy (1976), regard all effective groups as having three quintessential aspects to their processes — namely:

- a notion of membership and an appreciation of its consequences
- delineated roles of members
- shared norms

Without these groups cease to have cohesion. Broadly we use these three aspects as a framework in the following sections. Some of the aspects of norms — those of the whole class, and the teacher's role in establishing a non-threatening learning environment were explored in the last chapter. Here we have interpreted the term 'norms' as being about both the inter-relationships between group members (group dynamics) and shared understandings, — for example, of working practices — that develop within a group. In developing our framework around these three aspects, we have outlined four criteria:

- gender considerations — to allow the exchange of social constraints brought about by the different social experiences of boys and girls. At different times it is important to allow space for the development

of confidence between members of the same gender group, or exploit the variety of a mixed group;
- cultural and/or ethnic considerations — to allow all pupils access to the rich variety of cultural understandings which a balanced heterogeneous grouping can provide. Again, though, homogeneous groups can sometimes provide strength and support for a common viewpoint;
- skill considerations — the range of scientific skills of different individuals appropriate to the task in hand. Again there are benefits in having a mix of skills, and sometimes of having all members of a group at similar levels of competence;
- conceptual considerations — the range of ideas or scientific understandings which pupils have on the topic in hand brings variety, but again 'like minded' groups can also prosper.

We have not sought to be exhaustive in our criteria, there are many more which could have been included. In particular, though, there is no mention of friendship groupings. The main reason is that these tend to be a common and well-tried method of creating groups with which teachers are very familiar. They will know the circumstances and the classes in which friendship groupings are the best (or worst) way of managing the class. There are undoubted merits in encouraging pupils to work with their close allies since the process of team building is likely to be short circuited. Nevertheless, we feel that the disadvantages often outweigh the advantages. Most important, when working with friends youngsters are often trapped into stereotypic role relationships and have little opportunity to explore other ways of working, the established leader, for example, is always the leader. Thus we have concentrated on other criteria, being pragmatic in our choices, so that these are both small in number and can foster the main point of the exercise, to develop pupils' understandings of science.

Looking at these criteria, teachers working in single sex schools where the majority, or all, pupils are from a single ethnic group may well feel the first two are criteria to be ignored, since they lie beyond the scope of the individual to control. While this may be so for membership of groups, it is not so for the nature of the activities involved. Schools where pupils are drawn predominantly from one ethnic group or exclusively from one sex need to plan their activities carefully to compensate for the lack of variety available. Otherwise they lose the opportunity to explore and explain ideas in circumstances where they are open to the widest possible range of influences.

Using Our Criteria to Create Groups

Many primary pupils work in a variety of different constellations depending on the task in hand. Primary teachers are experienced in manipulating groups for different purposes and well versed in the problems of classroom management that such techniques can raise. In secondary schools, the large groupings — classes or forms — may differ as pupils move from room to room, particularly after the selection of options. Pupils may be set; put randomly into 'mixed ability' groupings; streamed; selected or segregated. And, of course, within these larger groups several possibilities can exist together. Our main point in this section is that when any *small group* of pupils is created by a teacher, within the larger class group, there must be a positive purpose to the grouping. This may differ on different occasions, and the purpose will then affect the 'life' of the group.

In creating groups, the choice of purpose is the first step. It is important to note that the *purpose* is not the same as the *activity* which the group will undertake. For example, a teacher may plan a piece of project work. She will then create groups to work on their chosen projects. In the first instance she will take into account gender and ethnicity and she may well decide that in this instance some groups should be balanced and others homogeneous. The reason for this choice should lie within the purpose of the work. Similarly, she will then consider the skills of different pupils in relation to the purpose and choose group members accordingly. If, as in this case, the purpose is a piece of project work, groups might be arranged with a variety of skills. Alternatively, she may decide the project requires pupils to share and communicate their scientific knowledge and thus prefer groups based on a variety of understanding and ideas. Finally, she will choose activities for the group which will contribute towards the purpose and decide what roles her pupils are to play. As fine tuning, she will look at the individuals in the group to determine whether the roles they usually adopt in friendship patterns will interfere with the purpose in mind and if so adjust group membership accordingly. However, a piece of project work is a long term grouping activity and there may be several substages. She may decide that in order to be successful, certain youngsters need to enhance specific skills and she would create, perhaps for just one short activity, groups which were homogeneous. That is, at the same level of skill, since the purpose would be to undertake direct skill teaching. Activities and roles would then be chosen accordingly.

In other words, although we see the creation of groups as an exercise primarily in creating a heterogeneous balance on a variety of criteria, there may well be purposes which the teacher wishes to achieve which require using the criteria in such a way as to have homogeneous groups in one or more respects.

To summarize then, the questions involved in creating a group might be something like this:

- what is the purpose in mind?
- what gender and ethnic groupings would best fulfil that purpose?
- what balance of skills do youngsters need in each group — a variety or similar levels?

- what level of concept development do youngsters need in the group, a variety of levels or a similar level?
- what activities will fulfil the purpose?
- what roles will individuals need to play to carry out the activities?
- what specific difficulties lie ahead because of the individuals put together, and what action might need to be taken?

Clearly, we believe that applying these sort of criteria maximizes the potential of groups to communicate scientific ideas and understandings. There may well be other advantages to be gained from applying the criteria but, for us, that is the major focus.

To Balance or Not?

The decision to balance or not, across all four of our criteria, is one based on the purposes the teacher has in mind. Here we set out the case for balance.

The central issues of gender and science are well known and the reasons for ensuring that groups have a balance of boys and girls (under many circumstances) have been explored quite frequently. The interested reader might try Kelly (1987), Ditchfield (1987a) and Watts and O'Brien (1989) for a backlist of references to follow the debate. For instance, making sure that boys and girls become used to working together on scientific tasks at an early age will go someway to mitigate the corrupt but widespread assumption that 'science is (only, mainly) for boys'. It will also give girls the opportunity to develop confidence in a wide variety of situations, not the least of which is to be able to explore and use apparatus with confidence in a peer setting. However, these are social reasons, often used to combat stereotyping and not, for us, the sole reason for group gender balance. Part of the constructivist philosophy we outlined earlier rests on the belief that the world experiences of different youngsters can be quite different in themselves. Youngsters' learning about the world is interpreted through their own framework of experiences. While experiences of physical and biological phenomena can be broadly similar (after all gravity affects everyone's mass in an identical fashion), the interpretation of those experiences happens via their own conceptual frameworks. From this stance, the world experiences of girls can be quite different from that of boys. It is self evident, for example, that role models they recognize, the social stereotyping to which they are subjected, means that both sexes have a great deal in common which is different from each other. Our argument is that there will be some aspects of science for which there will be significant similarities in the frameworks employed by girls to be markedly different to those used by boys. We believe that there is sufficient difference in their experiences to create a situation worth exploiting in a positive sense to encourage the developing of scientific ideas. Much the same can be said, too, of ethnicity. The differences in cultural understanding youngsters bring as a result of their experiences is a powerful stimulus to contribute to classroom work (see, for example, Gill and Levidow 1987, Ditchfield 1987b, Nixon and Watts 1989).

What of the arguments that:

> if I make boys and girls work together it causes more trouble than it's worth

> putting them in a group is one thing, but I have only two Afro-Caribbean boys in my class

or

> the girls always want to work together and if I separate them, they become very quiet and hardly contribute at all in the group they are in?

Clearly, the pragmatics of classroom life are important to those working in them, be they teachers or pupils. There is little point in applying our grouping criteria and thereby damaging the essential fabric of the learning environment. The classroom context is vital, as is the makeup of the whole-class group. We have met classes with whom some teaching approaches work very well and others where the same approach courts disaster. The decisions about grouping criteria must reflect the purpose teachers have in mind, including ensuring that the lesson progresses in comfort for all concerned. And, undaunted, there are times to take courage in hand and, as Brandes and Ginnes (1985) suggest, tackle even the most least likely groups. Who knows, they may even get to enjoy the experience. What, then, of the practicalities of establishing balance? Point one: it is best to start early in the year when the class is still unfamiliar with the approaches of a new teacher in (possibly) a new situation. In secondary education it is probably best to start off with younger pupils and develop the techniques as these pupils grow through the school. A teacher of 11 year-olds told us:

> When they first arrive in my class, we have a getting-to-know-each-other session. I tell them that this is in order to establish the kinds of skills that people have, and to get to know what they think they are good at. I also tell them that during the year they will have the opportunity to work with everyone else in the class, at some point, and learn from them. I stress the fact that everyone in the room has important skills in science even if, at this stage, they are not sure what science is or what those skills might be. Then I tell them the kinds of skills a good scientist needs. Finally I get them to work in pairs and fours sharing information about each other.

Communication and Groupwork

It is an approach which works well when pupils arrive at their first science lessons in their new secondary school, having come from a variety of different primary schools. It stresses the value of each individual's contribution and sets the expectation they will work in a variety of different — deliberate — groupings throughout the year. Activity 3.1 below is aimed at teachers of middle-year pupils (Year 6–8). It can be used with older pupils too, with some adaptation. For younger pupils (Key Stage 2, for example) Activity 3.3 might be more suitable.

ACTIVITY 3.1 (10 minutes): designed for assisting pupils to think about science skills or when working with a new class.

BEFORE THE LESSON
1. The object is to find a way of putting pupils into groups in a way that appears random to them. Prepare two packs of playing cards with different coloured or patterned backs. Take only the same suite(s) from each pack (diamonds for example), the total number of cards selected depends on the number of pupils in the class. You can now sort pupils into an infinite variety of random or pre-designed combinations. You can group pupils according to the pattern on the back of the card, the face of the card or different combinations of the faces. In our example we have used the random element to enable pupils to learn about each other, and a pre-designed element to ensure gender balance.
2. Prepare a list of scientific skills (best, perhaps on a pre-prepared OHP transparency) — National Curriculum New Attainment Target 1 provides an excellent basis for this. We have provided an example list given at the end of the activity.

DURING THE LESSON
1. As the pupils come into the room, hand the boys a playing card, face down, with one type of back (say, blue). Hand the girls a playing card with the other type of back (red), also face down. They are not to look at the other side yet.
2. Then tell pupils that you want them to share with each other 'something they think they are particularly good at'. It might be a skill from within or out of school. They will be working in this exercise with a partner chosen at random but, during the year to come, they will have the opportunity to learn something from working with everyone in the room.
3. Ask pupils to find the person in the room whose card has the same face as theirs — gender groups being equal, this will, of course, pair boys and girls. Each person has one minute to tell their partner about their particular skill, while you keep strict time and indicate when it is changeover time to the other person's turn.

4. When the pupils' two minutes are up, introduce the idea of scientific skills. Still sitting in their pairs, ask pupils to give you ideas for what kind of skills they think scientists need. Add these to your list without necessarily first showing the pupils your ideas.
5. Now ask them to discuss for two minutes, with their partner, how the information they shared before can help them to be good scientists (i.e., what their 'being good at' offers towards being good at science).
6. They must now change groups. You can organize this by asking for particular combinations of face cards to move together, depending on the size of groupings you wanted. This might be, for example, all the picture cards (a group of three boys and three girls) — all even numbers below six (again, three boys and three girls) and so on. You could push the point by asking them to get together with those face cards which made a particular combination of numbers — for example those that make seven, or twenty-two. It depends how much turmoil you can stand. They now have one minute each to tell the new group what skills their partner of the same face card brings to science. They then have another three minutes as a group to work out what skills the whole group could bring. Different combinations of groups will bring different skills, but one effect of this deliberate manipulation encourages the notion that girls' skills are of equal value, since every group will contain girls.

Some Example Scientific Skills

(We have tried to make these simple and use pupil user-friendly language. There is a more detailed list later in this chapter which you may want to use instead or as well.)

Observing
Measuring
Using scientific instruments (apparatus)
Describing what has been observed
Recording what has been found out
Interpreting scientific information such as charts
Interpreting findings from experiments
Stating ideas about what might happen (hypotheses)
Recognizing different variables
Constructing a fair test
Following instructions
Finding patterns in results from experiments
Planning and designing experiments
Making predictions

In Activity 3.1 the randomness of the first groupings allows a sense of excitement because it is unexpected, and the time limits allow little opportunity for the 'I don't want to work with her/him' syndrome to set in. The teacher's role is very much that of creating a sense of fun and mystery that makes the learning interesting. The follow-up can branch into a variety of different activities and indeed this introduction itself could continue much longer than it's designated ten minutes. For instance, pupils could be invited to list one important scientific skill belonging to pupils in the room so that when new activities are introduced in later lessons, both teacher and pupils can use the information to help build 'project teams' for experiments, or for providing 'expert advice' for other pupils. The list can be displayed on the wall, and a positive climate of contributions can be maintained which allows pupils to see that future groupings are based on the skills contributing to the task in hand. Activities such as this can be used to facilitate single gender groupings when, for instance, it is important to explore and highlight issues for which girls and boys have different opinions or ideas. Teaching some of the issues involved in New Attainment Target 3 (Earth and Environment) for example may well involve looking at a variety of viewpoints, as Activity 3.2 below illustrates.

ACTIVITY 3.2
CONSIDERING CURRENT CONCERNS ABOUT HUMAN ACTIVITY

BEFORE THE LESSON
1. Collect together material showing the viewpoints of opposing groups in a debate — SATIS material (ASE, 1989) is a good source. For example, the views of the 'Greenham Common Women' can be used to illustrate aspects of the 'pro' and 'anti-nuclear' debate. It is important that information like this is presented so that pupils can identify with some of the issues, and to ensure that available scientific evidence is part of the package of information. They should be helped to make informed judgments on the basis of scientific knowledge.
2. Design questions for pupils to discuss in groups, which focus on such issues as:

- why are some people opposed to nuclear weapons?
- how can their use be defended?
- what scientific evidence is there to support views that such weapons should be banned?
- would you agree that these issues are more important to women than to men?
- what scientific evidence might you have presented to persuade the Greenham Common Women to change their minds?

- what scientific evidence is there to support the views that nuclear weapons are important to the continuance of our society as we know it?

DURING THE LESSON
1. Group pupils in separate gender groups.
2. You must choose how to deploy the information packs: giving the boys the 'pro-nuclear weapons' information or (preferably) the information pertaining to the Greenham Common women. A further stage of the activity may be to have them research some of the scientific information to support their answers, rather than providing it for them.
3. Ask each group to discuss the questions and arrive at one group answer to each: it is a group discussion exercise, not individual comprehension. You might tell the groups that they can have particular pairs to answer particular questions when they return to share the answers with the whole class, but that final answers have to be agreed by all of the small group.
4. As a final part to the activity, ask each group to report their discussion allowing the audience to challenge only on the basis of scientific evidence — if they think that the arguments are not really scientific, or evidence, but simply supposition or opinion.
5. There are a variety of things to be achieved from a follow up — some empathy by both boys and girls for the women's position; issues of gender stereotyping; issues of scientific evidence and what constitutes validity. The activity itself could be cross-curricular between science, English, social studies and economics.

In Key Stage 1 some of the issues are different. In this case pupils ordinarily work in a variety of groups and there is less need to introduce the notion of randomness. It is still vital, though, that teachers ensure gender and ethnic balance in groups. Where groups are set by age, or development in particular areas then, statistically, a balance of gender and ethnic groups would evolve anyway. It is still important to guard against the potential for stereotyping, particularly for those pupils for whom English is not their community language. It is too easy to assume that language 'ability' is equal to the capacity to express themselves in English. Arguably, these pupils need a balanced grouping more than most. Key Stage 2 pupils commonly work in a variety of groups, though by this time much more will be known about their skills in science. More importantly, they will have begun to have firmly established friendship groups. It is important at this stage to begin to use 'random grouping techniques' to avoid the friendship trap. One straightforward way of doing this is to display a list of pupils at the beginning of the year, and mark each name with a shape or a number. As in grouping activities for older pupils, the scene needs to be set by telling the pupils that during the year they

Communication and Groupwork

will have the opportunity to work with everyone at some time. The list is then used to check this is happening and to direct pupils to different groups for different activities.

For example, groups can be changed with each new topic or theme. If the identification of pupils on the list is used carefully, it is easy to construct a balance of gender and ethnicity. There may be times, though, when forming the group can be part of the activity itself, for example in the Activity below the study of fruit provides its own mechanism for group selection.

ACTIVITY 3.3 for pupils in Key Stage 2

BEFORE THE LESSON
1. Collect a variety of fruit and vegetables, one piece for each group is sufficient. The fruit and vegetables must be from different countries and/or be part of the regular diet of different cultures. It is best to have a variety of colours, textures, shapes and sizes.
2. Write a word relating to the texture, colour, shape or size of the fruit on a series of cards or pieces of paper. If you use all four, this will permit groups of four.
3. Prepare some questions and activities featuring the fruit/vegetable. This can be a variety of things focusing on such issues as
- where does it commonly grow?
- in what recipes is it found?
- what does it look like inside?

Other activities can include drawing it, measuring its size, exploring its symmetry, finding out what kind of plant it grows on, looking at its seeds, trying to grow one themselves from a seed, making a print from it, using it to create a dye, etc.

DURING THE LESSON
1. Place a piece of fruit/vegetable in the working position for each group. This may mean more than one piece on each table, depending on the organization of your room.
2. Give each pupil one of the cards. Make your choices as you do so — for example, give all the girls a colour card and size card, all the boys a texture and shape card. Or alternatively, distribute the card by ethnic background. This is particularly helpful in assisting such pupils to answer questions on fruits/vegetables which relate to their own culture, since it gives them an area of expertise. The pupils read their card, or match shape or colour if you have drawn the vegetables to ease language skills.
3. Pupils walk around and place themselves in a group where the fruit/vegetable best fits with their word card. They must then explain to

> the others in their group why they chose this fruit/vegetable as being the most like their word.
>
> 4. Finally they are all required to compose a description of their fruit/vegetable which takes into account the four characteristics on which you have chosen to concentrate, and to discuss the questions or do the activities you have designed.

The grouping is built into the activity. It does, however, assume that initial class organization is whole-class based. This may well not be the case in all or indeed many) primary schools and it will be for teachers to decide which type of classroom organization suits best. Sufficient to say, issues of gender and ethnicity remain important and, in our view, should feature in every grouping decision.

Activity 3.3 fits well in a theme on food or nutrition and there could be a variety of follow-on activities. For example, parents could be invited with a group to cook a dish using their particular fruit/vegetable, and talk to the whole class about the fruit in their own country. Further work might include 'how long does it take to cook' or 'what happens to it when it is frozen then thawed'. The activity itself satisfies a great many of the observation requirements involved in New Attainment Target 1, particularly at levels 1 and 2. Issues such as recording findings, quantifying variables, or using a range of measuring instruments (as in NAT1 levels 3–5) can also be explored. Mathematical issues such as symmetry (Maths Attainment Target 10) can also be developed. These last three activities show ways in which grouping can be managed in order to enhance discussions. However, we are still only at the stage of simply effecting the groupings, not managing the activity to ensue that pupils do have the opportunity to share their ideas and so add to the richness of discussion available.

Taking into Account Skills when Creating Groups

Our other two grouping criteria are concerned with skills and ideas. There is no doubt that there are differences between pupils, in their experiences, speed of understanding, affinity for different styles of learning and teaching, and it is not our intention that such differences should be ignored. The criteria outlined above use skills and scientific ideas as positive attributes to be used. However, there is still the hurdle of which skills (and the level of those skills) pupils possess. Assessment for the National Curriculum may provide some of the answers: though continuous assessments — teachers' professional judgments watching pupils working — will undoubtedly remain the backbone of both formal and informal practices throughout.

We have in mind here only a coarse gradation of skill level: can they properly use a thermometer? a top-pan balance? read a Newtonmeter? install

Communication and Groupwork

A Skill Wheel

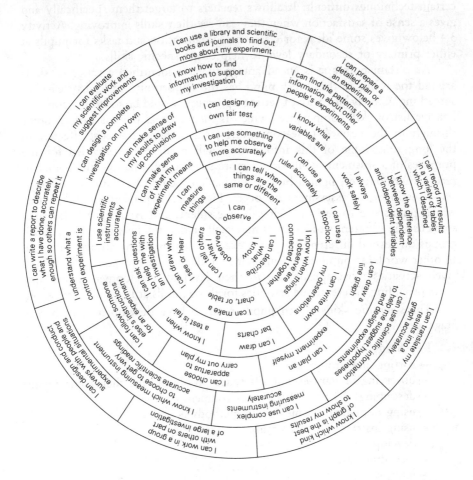

When you are sure you can do the things described in each statement, colour in the ones you can do and show your teacher.

Figure 3.1: A Skill Wheel

a voltmeter? read a universal indicator paper? set up a microscope? Establishing some basis like this will make grouping much easier.

The first possibility is to have the group at roughly the same level of ability to target that particular skill and provide enhancement activities. This is how groups in primary schools operate for reading and mathematics and so is a natural extension of existing practice. It is more difficult in science at the moment, though there are checklists and schemes available, particularly at secondary level and a variety of skill testing and teaching can be adapted to primary level. It is a useful approach for groups of pupils who are finding certain techniques difficult. It allows teachers to target them specifically and gives a sense of satisfaction when they can see their skills improving. Activity 3.4 below gives some ideas for classroom organization and tasks for pupils at either primary or secondary level to help target specific skills.

At certain points it is an advantage to have identified a range of skills around the room so that every pupil has the opportunity to lead or excel at something. Clearly one way of setting up groups would be to keep a comprehensive list of skill competences which is then used as the basis of grouping throughout the year. A time-consuming task, perhaps, but cost effective in the long run. Some of this information will need to be recorded for the purposes of assessment, while other, more specific information (the ability to read a thermometer, light a bunsen burner etc.) could be the basis of pupil self-assessment, or assessment by peers. Figure 3.1 represents a 'skill wheel' which pupils use to record their self-assessment, and Activity 3.5 describes some uses in different situations. It may well be that teachers wish to establish groups on several different bases, for example: grouping by attainment level. In other words, as pupils arrive in their first class of Key Stage 2, they will be given activities most appropriate to their level of development. This may mean providing differentiated activities for those pupils at Level 2, 3 and so on.

The scientific skills which pupils need to have developed at each Key Stage are:

Key Stage 1
— sorting and grouping
— observing similarities and differences in objects and their environment
— describing objects and their environment
— using non-standard measuring methods
— using instruments to measure
— developing ideas
— recording results
— interpreting results
— presenting results and ideas orally and in other ways

Key Stage 2
— observing objects and events accurately
— formulating testable hypotheses
— understanding notions of a fair test
— making decisions about the appropriateness of measurements and the form of them
— using instruments accurately for measurement
— manipulating key variables
— recording data systematically
— searching for and being able to identify patterns in data
— interpreting data and making changes to a procedure in the light of interpretations

- making written and oral reports on their findings
- developing a technical vocabulary and using it accurately
- using a computer to store, retrieve and present work.

Key Stage 3
- observing objects and events accurately
- formulating testable hypotheses
- planning investigations which incorporate the notion of a fair test
- understanding how to plan an experiment involving discrete and continuous variables
- manipulating a series of variables in one experiment
- accurately using instruments to record information and enhance observations
- systematically recording appropriate information about the activity
- interpreting collected data using mathematical relationships where appropriate
- searching for patterns in data and using them as the basis for making simple predictions which can be tested
- using technical vocabulary accurately
- using data and information stored on a computer as well as their own first hand observations as the basis for decision-making about an investigation

Key Stage 4
- being creative and inventive
- making accurate observations aided where appropriate by instruments
- planning activities in detail and evaluating the plan in the light of findings
- using a variety of secondary sources accurately
- dealing with investigations involving discrete, continuous and interacting variables
- controlling key variables and understanding the reasons for so doing
- generating theoretical models and testing them by investigation
- using sampling techniques in data collection
- measuring accurately using complex instruments where appropriate to the investigation and accounting for any anomalous results or experimental errors
- presenting results in a variety of ways and forms
- searching for patterns in complex data and making predictions which require abstract reasoning.
- critically evaluating data of their own and from secondary sources
- producing written critical accounts of investigations and ideas.

Within each of these there is a variety of shorter sub-skills involved. Observation, for example, might mean using eyes, hands or ears (or all three) and each of these would be a small subset to be checked during the course of an appropriate investigation. Using measuring instruments might mean being able to use a metre-rule, thermometer, ammeter or colour code accurately.

Most skill teaching in science occurs on a need-to-know basis — as and when pupils need it — commonly within the confines of a particular experiment. It is not usual to find a specific period of time devoted to learning that skill. The Graded Assessments in Science Project (GASP) (ILEA, 1987) suggest that skills are part of the work already being carried out, but that pupils can do extra work to achieve a 'skill assessment' mark, proving they have mastered the skill. This may well be an approach to adopt — there are several good ideas within the GASP project, as well as within the newsletters of the Science in Process Team (Heinemann/ILEA, 1987).

ACTIVITY 3.4
BEFORE THE LESSON

1. Prepare some skill exercises for pupils. These can be the kind of examples GASP or TAPS suggest, or ones that you make up yourself. You need to have them on hand for pupils throughout the year. The ones with stars against them in our list above lend themselves particularly well to this kind of activity.
2. Prepare a list of pupils' names as a wallchart, with the skills to be achieved throughout the year along the top of the list.

DURING ANY LESSON

1. When a lesson involves a particular skill — reading an ammeter, thermometer, using a balance etc. — if you are teaching in the whole class mode you will want to be sure that everyone understands how to perform the skill, by showing and telling them.
2. Similarly in a small group, you will want to check that pupils know what to do before moving on to other groups. However, it is difficult to check that each individual is performing the skill properly. Thus in order to be sure that each pupil can do what is needed, you can invite pupils throughout the next few lessons to try a skill exercise. This means drawing them to one side and asking them to work through the skill exercise for you. If they complete it successfully, and do some parts of it in front of you so that you can watch their technique, you can award them their skill badge or sticker and tick off their name on the list.
3. If you are working at the level of the group however, then there are several ways of arranging pupils' skill practice. First you can ask the group to decide when they are ready as a group to do the exercise and arrange it for them one after the other — or all at once if it is a simple one. This can have advantages in that the pupils are all doing the same

activity and you are not removing individuals who have an important task to do in the group.

4. Alternatively, you could suggest to groups that each person needs to take a turn at performing the skill in the group and that whilst everyone else is refining the skill, they need a resident 'expert'. Suggest that the group decide who is performing the skill best at first and nominate this person to do the exercise first and gain their skill badge. This person is then the 'expert in the group' and it is up to them to help others and, more importantly, ensure that everyone else does the skill exercise when the expert judges they are 'ready'.

5. For pupils who are having difficulty in mastering the skill, it is useful to use the method described in 4 above. This means that you can allow the pupil having difficulty to 'shadow' the 'expert' for a while and watch the skill performance at close quarters. It also means that you can maximize the peer tutoring capacity of the class.

6. The list of skills mastered is a very useful tool for teachers in making decisions about group composition. It enables you to choose groups that are well balanced and inform pupils that one of the other purposes of the lesson, as well as performing the experiment, is to ensure that everyone has the chance to learn something from everyone else.

7. The above descriptions work very well with pupils at Key Stages 1, 2, and to a large extent 3. In Key Stage 4, however, the notion of a skill badge or sticker may well not be an incentive to pupils. However, the mastery notion of practice followed by recognition as an expert who can help others, signalled on a list for peers to see, is still a good one. It works well and gives pupils confidence for GCSE practical assessments.

ACTIVITY 3.5
USING A SKILL WHEEL

1. The purpose of a skill wheel see (Figure 3.1) is to encourage pupils to develop and record their own self-assessment. As you can see, it mentions some of the things that pupils will need to do for New Attainment Target 1, but it also draws out basic measurement skills they will have met at different stages in order to fulfil some of the knowledge and understanding needs of the National Curriculum. It is not intended to be a self-assessment match for the National Curriculum. That is, the levels as they are in the wheel do NOT necessarily correspond to levels 1 to 10 in the National Curriculum.

Ways to use the skill wheel:
    The skill wheel can be used for pupils to record their own skills, as

they feel they have achieved something. This can be a record after the skill exercise has been completed similar to the one you are keeping on the wall.

It can be filled in as pupils feel they have mastered a particular skill, and the completion can then act as the signal for you to ask them to complete a skill exercise so that you are sure they have mastered the skill.

A group can use one as a record, filling in the relevant section only when everyone in the group has mastered the skill, or done their skill exercise.

A group can use the wheel as individuals. At the end of an experiment or piece of work they can look at the skill wheel, decide which skills they had to use in the experiment and then complete each other's by group decision. That is, the group decides who has mastered a skill — and had the chance to practise it and can thus complete the section. (This requires careful checking from time to time by the teacher to ensure that all group members are receiving positive rewards from their peers.) It is best not used when the group is one that has been in existence for a short while — perhaps the duration of one experiment. It is better used as a tool when a group has had time to establish itself, develop norms and the members trust one another.

It can be used by pupils to practise skills at home, with parents helping them to fill it in. This works very well in Key Stages 2 and 3, and as long as the instructions to parents are clear, can do much to assist parental understanding of work in science.

Using Pupils' Concepts and Ideas to Create Groups

We noted in Chapter 1 that pupils' concepts arise as a result of their experiences. Within any one class of pupils then, there is likely to be a range of conceptions about a particular scientific issue. At first sight this might appear to be problematic. After all, if one pupil's understandings are slightly different to those of others' and different again to the 'orthodox' text book explanation then 'putting them straight' may well be a problem.

Research also points to the fact that some well-worn individual conceptions are likely to be difficult to change, and there is little evidence that the traditional methods of 'putting them straight' are fully effective. However, small group work may well prove to be one of the more effective answers to the problem. If we are to encourage pupils to take on another point of view, or change their concepts then we need to ensure that they are exposed to a variety of understandings that will help them to re-examine their own and see whether it still 'fits' in the light of new evidence.

First, though, it is important to know what ideas pupils hold about a

Communication and Groupwork

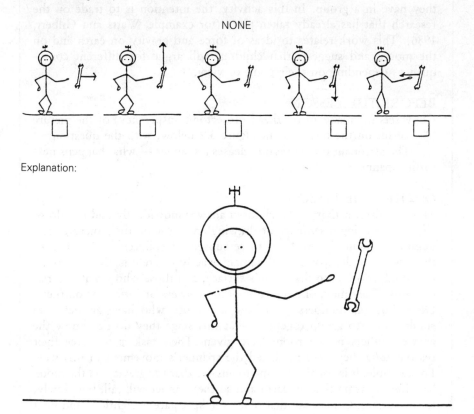

Figure 3.2: Astronaut on the moon workcard

particular topic before we can group them so that they meet opposing ideas. This requires some initial work which will act as an indicator for the teacher. The intention is not to conduct a battery of tests, but simply to gain some general information about where pupils are up to in their thinking. There have been several books published which contain a variety of drawings, experimental situations etc., which can be used in a variety of ways, and on a variety of topics.

ACTIVITY 3.6

There are many ways to pick out the particular ideas children have on a topic — from their written work, their homework, the questions they ask, the oral answers they give in class, or the kinds of conversations they have in a group. In this activity, the intention is to trade on the research that has already taken place (for example, Watts and Gilbert, 1986). This work relates to ideas of force and gravity on earth and on the moon, and suggests that children will argue for different consequences depending on their point of view.

BEFORE THE LESSON
Copy sets (or make an overhead projector transparency) of the picture 'Astronaut on the moon' as in Figure 3.2 below, with the question:
    The astronaut on the moon releases a spanner — what happens next to the spanner?

DURING THE LESSON
This is a short activity. The children answer individually and can do so either by writing a sentence or drawing an arrow on their picture. The question commonly produces three subsets of children in the class — those who say the spanner will remain exactly where it is, those who say it will fall (slowly) to the moon's surface, and those who say it will rise upwards. They may, of course, be other answers or variations on these. Depending on numbers, you now ask those who have answered in similar ways to group together — at this stage they do not know the answers others in the room have given. Their task is to agree their reasons *why* they have predicted the spanner's movement in this way. For example, it is usual for some to answer that the gravity on the moon 'is different' from that on earth and so the spanner will fall, but slowly, to the surface. Others say that objects in space are 'weightless' and float around, or that this happens because there is no air on the moon. A further set of responses tend to suggest that because the gravity is different only heavy objects will 'stay down' and that smaller ones like the spanner, will float away. Needless to say, there are other forms of response, but this kind of sorting — around the ideas children have — means that class activities can then be organized to tackle each of the points of view in order to examine their likelihood. Many of the references in the bibliography detail the sorts of ideas youngsters have for particular concepts in science so that activities like this can be planned for other topics.

Developing Particular Roles Within a Group

Here we explore:

- roles within activities for definite tasks, allowing particular contributions to be made;

- opportunities to practise new roles;
- group leadership and the effects of leadership styles on group performance.

There are two main ways of linking roles to the activity. First, for example, by assigning roles such as:

planner
reporter
data collector
experimenter
apparatus manager etc.

The roles need to be made explicit so that pupils know what is intended by each. An alternative is to take a 'process oriented' viewpoint and distribute the roles which occur in every group:

leader
worker
thinker
doer

Johnson (1985), for example, describes roles such as 'starter', 'encourager', and 'checker' being assigned to help groups get started. We tend to highlight the role of leader since it is crucial to the success of the group.

Fitting the Roles to the Activity

Each member of the group needs a role which contributes positively to the whole task, and this needs to be made explicit. Many secondary teachers commonly use a worksheet approach which outlines apparatus, method diagram, questions etc., and in these cases an extra heading could include the roles within the activity. An example is given in Activity 3.7 below, linked directly to Key Stages 2 and 3.

ACTIVITY 3.7

In this activity, the group of pupils are required to plan an experiment to investigate the way the temperature of a solvent and volume of the solute affect the solute's solubility. Pupils are given a list of useful apparatus including the solutes, and instructed to use water as the solvent. The activity fits well with work to illustrate aspects of New Attainment Target 4. An example of a worksheet is given below.

BEFORE THE LESSON
1. Look at the activity, and decide briefly for yourself how you think pupils might carry it out. Based on this, choose roles you want in the group which are task related. This decision will affect the number of members of each group.
2. Decide who is to be group leader and whether this role is going to be separate from other task-associated roles.
3. Based on your decision in 2 above, you may need to write some guidance for the group leader, telling her how to conduct the group and what is expected of her.

DURING THE LESSON
Our decision was to allocate roles as described in this example worksheet:

In your groups you will need to plan how to do an experiment to find out about how things dissolve. You will be given a list of the solids (dissolving things, called SOLUTES) and you should use water as the liquid in which they are dissolving (this liquid is called the SOLVENT). At the end of your planned experiment, your group will need to make a report. The report should tell everyone else what you have learnt about:
a. the way in which the temperature of the SOLVENT (water) affects the amount of solid which will dissolve
b. how much SOLUTE (solids) can be dissolved in the same amount of water if the temperature of the water stays the same all the time.
c. what you think has happened to the SOLUTE when it dissolves.
There is a planner sheet to help you plan out the stages of your experiment.
In planning and doing this experiment, there are several jobs to be done. This means that each person in the group will be responsible for doing different parts of the experiment. Each person will have what is called a *role* to play in the group. The roles needed for this experiment are:
1. *Group leader.* This person helps the group to decide on its plan, makes sure that they keep to time, and that all the things needed by the experiment are completed. She/he keeps the group working well together and makes sure that any gaps are filled — that no jobs are left undone.
2. *Recorder.* This person is responsible for writing down the decisions the group makes, filling in any sheets, like the 'planner sheet' for example, and helping the group to write a final report together. She/he does not write the group's report themselves, she/he helps everyone to work at it together, and record what the whole group decides. She/he is also responsible for keeping accurate results of the experiment.
3. *Reporter.* This person reports what is going on. If the group decides to split up into two sub groups, this person has to keep both sub groups

> informed of progress in the other group. She/he has to make sure also that everyone understands what is going on, and why different parts of the experiment are being done as they are. If there is a verbal report-back to the rest of the class, this person should do it.
> 4. *Technical organizer.* This person is in charge of organizing the actual experiment. She/he has to give each person in the group a job to do in the experiment. Some jobs will be part of a person's role anyway, such as the recorder, who has to write down the results. But there are also many other jobs, like collecting apparatus, conducting a part of the experiment, measuring things, labelling things etc.
> Decide who is playing which role.
> Now on your planner sheet, fill in the names of the roles and the persons doing each one.
> You can now begin planning your experiment.

Leadership

Group leadership is important. One dilemma facing any small group, particularly of younger pupils, is how to deal with leadership style. The first decision is: should the leader be teacher-selected or allowed to arise naturally from the interplay in the group? Unabashed, we favour teacher selection, most of the time. We want pupils to develop experience of a variety of roles and these need to be allocated in a planned way. Of course, the underlying rationale for all of these decisions is the purpose of the grouping in the first place. So one may decide to rotate leadership in long-term groups, but allow leaders to arise naturally when groups last only for the duration of one experiment.

Adhering to roles is important, in some circumstances it is not uncommon for groups to develop an 'alternative leader'. This can interfere with work and progress since members are uncertain where decision-making really lies. Dissension and indecision can result in inaction.

What Styles of Leadership are Possible?

The best known research on leadership styles is that of Lippitt and White (1960). They outline three major types:

- Authoritarian — characterized by policy formation by the leader; action determined by the leader and revealed a little at a time; dictating tasks to the rest of the group.
- Democratic — characterized by policy formation by group discussion and decision; individual activities arising through discussion; the leader outlining alternative strategies from which the group choose; the division of tasks left to group members;

- Laissez Faire — characterized by complete freedom for policy making; a minimum of leader participation; action decided by the group, the leader supplying only information and taking no part in discussions; tasks carried out without involving the leader.

They indicate quite different leadership styles and effectiveness: groups with Authoritarian leaders spent the most time on the task, but the work rate dropped with the leader not present. They also displayed much more aggressive behaviour. The Democratic leaders' groups spent half their time on task, but produced work of a superior quality and were most involved in what they were doing. The group could also work for half the time very productively without the leader present. The Laissez Faire groups spent a third of their time on the task, a third playing, and only 10 per cent working in the absence of the leader.

Clearly leaders are important in determining how much work is done and how much satisfaction group members experience. Further studies by Maier (1963) tend to indicate that strong leadership, of the more authoritarian style was most effective in bringing a group towards solving a problem. Moreover, to avoid the laissez faire, we need to teach youngsters how to balance the productive aggression of the authoritarian style with the relaxed devolution of the democratic. More recently Houston (1984) has outlined four types of leadership, in a humorous but recognizable way, focusing much more on the emotional and personality factors of leadership, 'pen portraits' which are very useful in helping pupils understand and recognize such leaders:

Houston's Leaders

- The 'wonder child' is all things to all people. Does all the organizational tasks herself, understands everyone's problems, has opinions on everything and gives no space for the abilities of group members.
- The 'loving shepherd' radiates charisma but cannot back words with deeds. Such a person basks in warm feelings and manipulates others into doing the bulk of the work.
- 'Idle Julie' sits quiet much of the time and tries to analyze what is going on in terms of the forces in the group.
- 'Just One of the Group' type of leaders tries to model full and honest communication. She might develop guide-lines for the group but she sees her role primarily as a none directive one, not a leadership one.

What Does This Mean for Science?

In a short term piece of work, such as an experiment that lasts for a lesson or couple of lessons, the real overall leader is the teacher. In most situations of

Communication and Groupwork

this kind, the teacher acts as an Authoritarian. She decides what the task is (policy formation) allocates jobs to be done and how they are to be completed. She then hands over authority for 'keeping on task' to the group leader. In an experiment which uses a worksheet, much of the decision-making power is removed from the group leader. However, within the framework of carrying out the experiment, it is possible to allow group leaders to develop some of their own style. They can become the spokesperson for the teacher and act in an Authoritarian mode, by allocating tasks to different individuals and making sure that these are carried out. The leader might also be encouraged to be more democratic in her/his leadership, by seeing the worksheet as a brief which has to be carried out, but allowing discussion of roles and tasks and who is to complete them. In other words enabling the group members to have some say in the procedures for the experiment. Pupils who find group leadership most difficult may well opt for a *laissez faire* style, since it puts them under the least pressure. It is obvious that pupils need to be able to understand ideas about how leaders can operate. This speaks of some direct teaching about groups. Activity 3.8 below may assist in this aspect. It is clearly a cross-curricular issue of skills which pupils need to enhance their learning in different subject areas.

ACTIVITY 3.8

Teaching the class about leadership.
The idea of group leaders needs to be introduced, with an activity based on Houston's leadership categories. This is followed up by discussion of how different leaders in the class operate.

1. Ask pupils to work in fours. Any combination will do here, it is a very short lived exercise and building the group is not necessary.
2. Ask them to think of particular situations in the playground or outside school where they have experienced someone acting as the leader of the group of friends they were with. Remind them that it should not be a situation where the leader has been chosen, but where they have followed someone's lead naturally. Request that they brainstorm for two minutes the kind of things the leader did which showed that they *were* the leader. You might need to give a couple of examples here such as — 'always had the best ideas' — 'was full of energy' — 'got everyone else enthusiastic etc.'. You will need to timekeep carefully and remind them that discussion is not permitted during the brainstorm. The object is to get as many ideas as possible down.
3. You might want to have some feedback from each group, but it is not necessary. The purpose of the brainstorm is to get them thinking for themselves about the kinds of characteristics leaders display.

4. Now give them Houston's leadership categories and the accompanying 'case studies' below. Ask them to decide which type of leader the case studies are portraying. Remind them this is a group exercise. They have to agree about their answers and present one answer from the whole group. This should take about ten minutes.

5. Take feedback from the pupils. Case studies 1 and 2, should be straightforward, but 3 may be more difficult because it is a hybrid. Use this to point out that not everyone does things 'by the book'.

CASE STUDIES

1. John is leading a group in a science lesson. He gets everyone to sit down, makes sure they have the instructions to read and can understand them. Whilst they are reading, he rushes off and gets the apparatus. When he comes back he asks the group if everyone is clear what has to be done and then gives everyone a job to do. Susan doesn't want to do what he asks her to do. She complains. John listens and then asks Paul to swop with her. He does. Whilst everyone is doing their bit of the experiment, John hurries round, checking readings, helping, adjusting apparatus. At the end of the experiment, he helps to clean up and then gives the report back to the teacher himself.

2. Everyone likes Surrinder. She is a very generous person who is very friendly. When the teacher asks her to lead a group in the Home Economics lesson everyone wants to be in her group. She gets the group started by ensuring that everyone knows what to do, and has volunteered for a task. They all collect the food and the cooking utensils. Surrinder sits with the food to make sure no one borrows anything from their table. She helps everyone to get organized, smiling and chatting, checks the oven is on, while Harry does the weighing, Kim makes the pastry and Stephanie fills the pie. Surrinder tells everyone she will wash up. Just before the end of the lesson she hasn't finished writing up her notes. She appeals to Wayne to do the washing up for her, telling him that she'll do it next time. At the end of the lesson she helps Harry and Kim put the pots away.

3. David likes to make sure everyone takes part in the group. He starts off his time as the leader in groups in French lessons by negotiating the rules with everyone — do they want to work in pairs or as a whole group for example. He always does what the group wants, but once they have begun work he joins in enthusiastically. He is full of energy, dashing about and helping everyone as well as getting his own tasks done. At the end of the lesson, however, he remains quiet, making sure someone else from the group reports back.

As we pointed out within the activity, in considering the role played by the leader, it is important not to forget that a group is made up of several individuals who interact and that knowledge and understanding of what is happening in group processes is very important.

Norms and Group Dynamics

The issue of norms is an interesting one. They refer to the norms of the large group — or whole class — as much to the effective working of a small group. As we explored in Chapter 2, the teacher has a vital role to play in establishing the norms of expected behaviour, the non-threatening learning environment. The establishment of such an environment is essential before any successful group work can proceed. For example, to facilitate small group work, one crucial 'norm' which the teacher must establish for our grouping criteria to work is the expectation by all pupils that they will, over any one year, be part of several groups for different lengths of time but they will all have things to contribute and things to learn in every group. Once that 'norm' has been established and groups organized based on the idea of encouraging skill exchange and learning, then the work of small groups can proceed towards making their own shared understandings.

We referred in Chapter 2, to Hart's (1989) research. As he pointed out, a crucial part of the teacher's role in ensuring that the dynamics of the small group work, lies in setting the parameters within which pupils in the groups can negotiate their working practices. It is one way of keeping these established norms of a new threatening learning environment explicit and giving pupils the opportunity to negotiate them.

ACTIVITY 3.9
DEVELOPING A SCIENCE CLASSROOM BEHAVIOUR POLICY FOR MIDDLE-YEARS PUPILS

BEFORE THE LESSON
It may help to have some of the Science in Process material about safety available to prompt some pupils for this lesson. This is composed of pictures of activities in the laboratory that are unsafe. It could also form a useful follow-up exercise, such as asking pupils to design and draw their own posters for their laboratory.

DURING THE LESSON
1. Divide pupils into fours.
2. Ask them to brainstorm for two minutes all the actions they think are unsafe in an area where science is taking place.

> 3. Ask them then to look through the list, delete any duplications, then decide *why* they think some of the actions are unsafe.
> 4. Gather the whole class together for feedback. Take one response from each group, until there is a complete list — no duplication is allowed. What is most likely to emerge is a list of negatives ... Don't run, chew, etc.
> 5. Now discuss with the pupils how these negative statements can be turned into positive actions so that they feel good about the way they behave around science areas and not just negative.
> 6. When some agreement has been reached, ask everyone to write their own 'Behaviour Policy'. Put a copy on the wall in chart form for pupils to refer to.
>
> You may want to revisit it from time to time, to see if it needs alteration in any way. You may also want to appoint a pupil to be safety monitor for a week using the policy as a guide so he/she can tell the class which of the policy actions most people obey and those which most people ignore.

There are obviously aspects of behaviour, as we explore in detail in Chapter 2, that apply both at whole-class and small-group level. Undoubtedly small groups will establish their own norms too, but the establishment of an NTLE is essential in helping the small group to become established and work effectively. As Bill Rice (1981) points out: 'A group which fails to unite (or gel) properly may prove unable to undertake successfully collaborative (learning) ... The degree to which a classroom group develops is influenced by the willingness or otherwise of individual members to identify with the group' (our parentheses).

He continues, 'The emphasis should be on cooperative work rather than in promoting competition between group members or other groups. Collaborative activities tend to emphasize that groups should hang together' (Rice, 1981).

Class and Small Group Rules

If the non-threatening environment is to be successful in the small groups, to encourage the collaboration which Rice, Yeomans, and Hart (op. cit.) are advocating, then the whole class norms will need some reinforcement. It is easier for a teacher to remind pupils about behaviour which transgresses the class norms when the grouping is the whole class. It is more difficult when there are several groups operating in a room, perhaps all engaged on different activities. The work in small groups may need some supporting structures so that everyone can use them as a basis for working and be protected by them, with a right to challenge those who infringe them, including the teacher. Some rules which have been found to be useful and refer directly to discussion in small groups, are:

- listen carefully to others and give them time to put their argument fully to the group;
  Interruptions before they have finished explaining are not acceptable;
- always try to be succinct. There is a limited amount of time available for discussion. Everyone deserves the opportunity to share some of their thoughts. Give them this chance by being brief and sticking to the point;
- if criticism is necessary, criticize the argument or idea, not the person who presented it;
- ask for explanations if you do not understand what is being said. The discussion is always improved if everyone understands. It also helps the person who is explaining to understand ideas better themselves if they have to find another way of saying it to help someone understand;
- the whole group will benefit if you ask questions about the ideas such as 'what do you think would happen if ...?';
- try to watch out for those group members who find speaking out difficult. They often have very good ideas but need encouragement to share them. With the benefit of those ideas, the work of the group could be that much better. Give them support and encouragement and everyone will gain.

Getting pupils to work to these 'rules' or establish some for themselves will ensure that the class norms are carried successfully into the small groups.

Working Practices

Every group needs to establish its own working practices within the agreed norms. The 'rules about discussion' for example will ensure that respect for other's ideas are foremost in the work of the group, but one should not assume that this will carry over into working practices in experimentation. Even with established roles of reporter, manager etc. in a group, some groups still manage to work in totally different ways. They establish sets of ground rules for themselves that feel most comfortable for their individual learning styles. For example, some groups will prefer to work as a loose collection of individuals who come together only when each has accomplished their own part of the overall task. Others very quickly move towards a collaborative approach, and if they work together regularly, begin to carry out their normally allocated tasks without having to refer to the rest of the group. Such groups are usually efficient, use time very economically and work towards a common purpose. They seem to develop a 'subliminal understanding' of the norms that hold the group together and work towards developing rather than transgressing these. Partly, such understandings come from knowing how others in the group like to work, and the kind of skills and expertise they have. However, although some groups, perhaps with particularly sensitive individuals in them, seem to develop such understandings implicitly for themselves, other groups can be helped along the road to such understandings by providing them with opportunities to stand back from the task and think about what is happening in the working of the group. In keeping with our previous activities on working with the group, skill teaching activities, below is an exercise designed to help pupils understand how groups work. That is, reflect the stages shown below.

Establishing and Using the Dynamics of Groups

In this section, we spend a little time examining the ways in which individuals interact to form the group atmosphere in small groups. We feel this is an important aspect to understand if we are going to help groups proceed towards successful task completion. In terms of the views of learning which we expressed in the first chapter, if communication is to be enhanced by group processes then paying careful attention to the group dynamics is important. As Yeomans states 'the ability to manipulate *class* group dynamics is a fundamental teaching skill' (Yeomans, 1988 p. 163). However, organizing the dynamics of a small group is different from that of organizing those of the class. Controlling small group dynamics may well be difficult when a teacher is rarely in contact with the group long enough to do what Emery (1988) calls 'listen to the music of the group'. That is, understand the feelings and atmosphere which often comes from the way in which things are said or arise because of the particular issues on which people choose to focus. For example, work is needed to help groups establish their own sense of identity and working practices, which are quite different from those of any other in the room. In many cases this comes only after the group has been established for a period of time and has had the opportunity to develop and understand quite what values and expectations are shared.

ACTIVITY 3.10
LEARNING ABOUT GROUP DYNAMICS

In this activity, it is necessary to use some pupils as observers during a science practical to focus on the dynamics of the group and reflect to the group what is happening. It is not an activity which needs to take place very often, but it can be useful to collect information from pupils from time to time on how they feel the group is progressing.

1. Ask one member of each group to act as observer. Whilst the rest of the group collect the apparatus for the work, collect the observers together and give them the instructions below. Tell them that they are responsible for collecting information and reporting back on how the group works as a group at the end of the lesson. They are not involved in carrying out the experiment.
2. Take feedback so that you can highlight the different features of leadership, what helps the group to work best, what prevents the group from working etc. There will be different answers as to what helps and what hinders from each group. This is a good opportunity to remind pupils that group dynamics depend on the individuals in the group and the interaction between them.

INFORMATION FOR OBSERVERS
Your job is to observe the way the group works.
At the end of the experiment your teacher will ask you to comment on the following headings. Each heading has some questions underneath. They are suggestions only. You do not need to answer them all, or at all. They are just ideas for the kind of things you might look for.

1. Is the group working together successfully?
You might look at such things as: do they finish their task, do they quarrel? Does everyone join in and do a fair share of the work? Are some people lazier than others? Does everyone do everything, or do people specialize in certain things? If so why?
2. What helped the group to stay on task or move in a new direction? Was it, for example, someone who kept introducing new ideas? Was this always helpful? Did someone keep disagreeing all the time and stop things getting done? Did someone have a really good idea that changed everyone's thinking?
3. Did everyone get treated equally?
Did boys and girls suggest the same number of ideas? Did they do the same kind of jobs? Was there someone who was very quiet or was never listened to when they suggested things? Why do you think this happened?
4. How does the leader react?
Are they very directive telling everyone what to do — a Wonder Child type — or getting everyone to join in and not really leading at all — a Just one of the Group Type? Or do they have a different style altogether?

Chapter 4

National Curriculum Tasks in the Classroom

This chapter represents a drawing together. In Chapters 1 to 3 we discussed the various communication skills that might be developed, explored ideas about non-threatening learning environments, developed activities about grouping, roles and norms. In this chapter we pull all these aspects together within the National Curriculum. The chapter is a series of tasks which teachers from each key stage might try. The tasks build on the ideas developed in other chapters, but put them all together to meet the requirements of particular 'acts of communication' as defined within the programmes of study. We have not tried to develop tasks for all of them partly because to do so would be repetitive in many cases, partly also because some of them occur only once in each statement, or are so self-evident as to need no explanation. We have chosen to develop the nine most commonly occurring acts of communication in the National Curriculum as tasks which teachers might want to use in developing communication in science. In developing the tasks, we have sometimes returned to the 1989 wording where particular attainment targets require a particular communication skills at particular levels. We have developed tasks suitable for the relevant key stages.

COMMUNICATION SKILLS	TASKS DEVELOPED FOR Key Stage
describe	2
	1
give an explanation	4
	2
keep a diary	1
give an account	2
evaluate	4
discuss	3
support views	3

TASKS FOR KEY STAGE 1 ACTIVITIES
ACTIVITY 4.1

COMMUNICATION SKILL: KEEPING A DIARY

We have interpreted this fairly liberally, since there are several occasions in the programmes of study when the recording of successive events is required to demonstrate change over time. For example, recording the different phases of the moon over a month's period. This could be done by keeping a diary and thus we have made reference to this in the activity.

NAT2 Level 2 entails being able to keep a record in a variety of forms of change over time. Our example here, as we refer to the programme of study, is clearly one in which the decay of waste is the change being monitored by keeping a diary.

BEFORE THE TASK

You need to decide what form the diaries are going to take. If you decide on a variety of records, such as, verbal descriptions, drawings, written descriptions, photographs, you could select different groups of pupils to use different forms of diary. One group does the verbal description and records it on tape, one the photographs and mounts them, one the drawings, etc. Having made the decisions, and collected the relevant apparatus, such as the school camera — or better still a Polaroid — (if a Polaroid is used, it can be operated easily even by six year olds, and provides 'instant' evidence.), a tape recorder etc., you need to introduce the activity that the pupils are going to monitor.

DURING THE TASK

A starting point

Since this task is a long-term one, the introduction and setting up can be brief. It may involve talking about the litter pupils see in the streets and asking them what they think happens to the litter that does not get put in a bin. Collecting pupils' ideas on a large sheet, to remind them over time of their hypotheses, is a useful way of recording. This may then be followed by a 'litter collection' and classification exercise, to see what types of litter pupils can find. For long-term experimentation, it is important to classify these items carefully in order to draw out notions of bio-degradability and non-biodegradability. It would also help to set up a 'Litter Area/Corner' where the long-term experimentation is to be stored so that pupils do not to have to dig up their buried exhibits from outside so often. The first part of the diary begins before the litter

> is buried. Each group needs to record what their litter looks like. This can be by each of the methods described above: some groups can draw it, or photograph it; others can write a description and some can record their ideas on tape. It is of course important that pupils date their observations.
>
> Discussion can then take place about how the litter is to be buried. Outside in the soil is an obvious place and closest to reality for pupils. However, this makes regular re-examination difficult and at the mercy of the weather. All the litter could be buried in a large glass tank, try borrowing one from the local secondary school, since they often have fish tanks that have begun to leak. These would serve excellently for this purpose. Alternatively, see-through biscuit boxes work just as well; the litter needs to be buried so that part of it can be seen. The rest is just a matter of adding to the diary records over a period of time.
>
> Other starting points
>
> Another approach to setting up the experiment would be to take pupils to visit a local refuse dump so that they can see the litter being piled, and ask the questions about what happens to it over time.
>   Alternatively, a history theme which involved archeological investigations could deal with the issue of what survives (and what does not) from previous ages to help us learn about their life styles. Visits to museums and examples of archeological digs would enhance this.
>
> Other suggestions. There are several aspects of long-term recording which lend themselves to this technique, some which one specifically mentioned elsewhere in Key Stage 1. One good example is that of recording the phases of the moon, as required in NAT3. The pupils could use similar methods for this (excluding, perhaps, the polaroid camera) and displays of their words and drawings — for example a mobile of different phases of the moon — would make an excellent display. Poetry, too, would lend itself very well to this activity, especially in terms of accuracy of observation. Photographs of the moon from space would also provoke debate about phases of the moon and why the space photographs always show the moon as round. Crescent shapes in mathematics could also be developed here.

It can be seen from Task 4.1, that, as far as science is concerned, pupils are using observational skills, learning about different kinds of recording and different types of evidence collection. In terms of the language development in National Curriculum English, the task itself gives pupils the opportunity to experience the Attainment Targets concerned with speaking and listening at levels 1 and 2, the Attainment Targets on writing, at levels 1 and 2, and Attainment Targets on handwriting and spelling. Follow-up activities could take place involving poetry, or a small playlet. The approach clearly lends itself to a variety of other aspects of science where long-term experimentation is required.

With older pupils, it is interesting to read to them diaries of scientists themselves, such as the ship's log kept by Darwin, or Milikan's diary, to provoke debate about methods of recording.

TASK 4.2
COMMUNICATION SKILL: DESCRIBING
NEW ATTAINMENT TARGET 2

This communication skill occurs in many attainment targets and the suggestions made here apply to a variety of other situations, although in this example we have limited it to NAT2.

NAT2 requires that pupils should understand the basic life processes of human beings. We have interpreted the programme of study to mean that children should consider the foods they eat and why they eat them. They should talk, or communicate by other means about what they eat, and why and when they eat.

In the last task (4.1) pupils worked in groups, though they could just as well have worked individually on the tasks set. Here, we show how group work can be used to enhance the communication of pupils about food, and when and why they eat. As with the last task, there are a variety of starting points:

Divide the class into groups of five or six using the criteria we developed earlier:
— Groups mixed for ethnicity if possible
— A mixture of verbal reporting skills and confidence in speaking
— A mixture of spelling ability
— A mixture of drawing skills

Start each group by asking them to keep a log of what they eat during one day. Ask them to bring in an example of some of that food — e.g., a vegetable — to show their group the next day. During the next day, bring together the group so that they can tell you, (or the parent or welfare assistant who is helping), what they ate the previous day. If the pupils are slightly older (6+) then maths activities (such as a histogram of the most popular foods eaten in one day) are possible after this task. In the group encourage pupils to listen actively. Ask them before they start listening to think about one food, mentioned by a group member that they have never tasted, or one that they like very much. Whilst they say what they ate the previous day, note the range of foods at each meal.

Read this back to the pupils and ask if they can see differences in what is eaten at different times of the day. Ask why. This is the point at which some issues of energy and food can be introduced. Ask each group member in turn to tell a neighbour which food they have *never* tasted or which they liked best. If there are some unusual foods mentioned by some pupils, try to draw out descriptions of the taste of the food, or what they most like about its the taste.

This task then continues with a description of the vegetables that group members have brought in. These could be placed on display for the whole class. One interesting way of doing this is to ask each pupil to describe the vegetable they bought, other pupils draw the vegetable as they listen to it being described. Obviously the pupils doing the describing should not name the vegetable! Within this there are a great many language issues which can be explored, and this makes it essential that all the groups have time with the teacher.

One follow-up activity is to ask pupils to taste a particular vegetable or food in order to demonstrate language issues and make comparisons between tastes and observations. Our example is drawn from cheese. You will need to supply a range of cheeses, full fat, semi-skimmed and skimmed milk for each group. Ask each pupil to taste a tiny piece of each cheese and describe the taste. Try to draw out words like 'strong' and 'mild' Ask them to rub each one on a piece of paper and describe what happens. Pour a little of each of the types of milk into a glass container and ask pupils what differences in appearance can be seen. Introduce the notion of fat in milk and cheese, and invite pupils to infer from the marks on the paper which cheese has the most fat. Show them pictures of polar bears, sealions and seals, and discuss why these animals must have a lot of fat under their skins and how it helps their survival.

This task explores the notions of grouping — around skills, and gender. At this age, work on developing roles and specific aspects of communication skills other than that of describing are not worked through. The task is designed to develop language skills and explore with pupils the meanings of words as well as classifying foods in different ways, such as 'foods I have never eaten', 'foods I like best', 'foods I eat at every meal', 'foods I eat only at breakfast' and so on. The need for small groups is obvious here. The pupils need time with an adult to explore their ideas and develop their vocabulary. Classroom organization would need to reflect this requirement. It is not expected that some parts of the task would take place with all the groups in the class doing the same thing at the same time. Rather, the work might be staggered across a day, so that different groups could hold their discussions with the teacher whilst other groups were completing their drawings, or doing other activities set for the day.

TASKS FOR KEY STAGE 2
TASK 4.3
COMMUNICATION SKILL: GIVE AN ACCOUNT
NEW ATTAINMENT TARGET 2

NAT2 states that 'pupils should know that human activity may produce changes in the environment'. The relevant programmes of study refer to children studying some aspect of the local environment which has been affected by human activity. Clearly the environment to be studied will be affected by local conditions. One project we know had youngsters who were involved in cleaning part of their local river and stocking it with fish. The project was clearly long term, so we describe the major stages and the accounts pupils gave at the end.

The groups are designed to ensure a balance of gender and (where possible) ethnicity. Skills to be taken into account include those of measurement, organization and planning, and recording or drawing. The roles necessary for a long-term project would be group leader, reporter, recorder and a technical organizer.

The task begins with a description by the teacher of different river conditions in separate areas of the country, perhaps with slides, or newspaper cuttings illustrating the problems of pollution. The class is invited to decide, in their groups, which aspects of pollution in their local river they will investigate. They make their choices and the next few sessions then involve visits to the river site, collection of data, and investigations of the data back in school.

Groups may choose to investigate a variety of aspects of the river, such as the types of different animals present in the water as an indication of the pollution, the number of fish present in a given volume, the amount of algae and plant life present in a given volume etc. They may also choose to investigate several different sites on the river to make comparisons.

Having chosen their area of work, groups should then be allocating tasks to each member, based upon the roles they have already been given. The final part of this series of lessons is giving an account of their work. The example we show here is where an assembly was used for each class to share its work, once a term.

1. Preparing the report.
Each member of the group is asked to write (or prepare in some other way) their own part of the report ready for group discussion.
2. When the group meets for its discussion, each member is asked to look at the parts together and then answer these questions:

— does it tell the story of all the things we did?
— does it tell everyone else what we found out from our investigations?
— does it tell everyone what we learnt?
— does it look attractive — are there drawings and writing?
— is all the writing in sentences? Must it be? Can some bits be in poetry, or could we have music for some?

3. As a result of their discussions, they should be given time to change their work so that the account they present is attractive, visual and stimulating for others.

4. Finally, they should decide who is going to make the actual presentation of their work and what is to be said, as well as who is to display other parts, sing, read the poetry, etc.

Time will obviously be a constraint, but the object is to remind pupils that accounts of experimentation presented for an audience should be stimulating, short, focus on the important facts and be visually informative.

This task is designed to provide the opportunity to cover several of the aspects mentioned in the last chapter. Some of the grouping activities or skill, or group dynamic activities could well be introduced as parts of the work during this investigation. Because it is a longer-term one, the task requires practising a variety of communication skills; reporting back, taking account of the audience, drawing, paraphrasing, etc. Within the groups there is the opportunity for working practices to arise which can be observed and commented upon to assist the development of communication between members and lead to a sense of responsibility for others. Assessment techniques such as those involving skill wheels could also be used from time to time, throughout the duration of the work. The activities which could add to the process and enhance some of the aspects discussed in the last chapter are: 3.5, 3.7, 3.10.

TASK 4.4
COMMUNICATION SKILL: DESCRIBE
NEW ATTAINMENT TARGET 2

The programme of study at Key Stage 2 asks that pupils are able to know the basic life processes common to humans and other living things. In this task, we focus on descriptions at level 3.

Obviously, in order to enable pupils to know and describe a process, there are a variety of teaching issues to be addressed first. For example pupils need to be introduced to ideas of flowers, their appearance, structure and function. This can be done by means of a variety of approaches, involving whole class teaching, for example, a video which focuses on ideas of pollination. It will also involve pupils working in groups with specimens of flowers available, describing what they see to others in their group and drawing what they see. They could be asked to speculate about the purposes of pollen and asked why the stigma is sticky at different times. In other words, it is necessary to have undertaken a fair amount of introductory teaching before the pupils reach the stage upon which this task focuses on.

BEFORE THE TASK
1. Prepare some sheets like the ones below to help pupils through the task.
2. Have available a tape recorder.
3. The focus of the description is by means of a visual playlet or narrated mime, in which pupils play the parts of the flower, the pollination agent and the final growing fruit. It may be best to suggest to the group that there is a narrator to help in the description of what is happening.

DURING THE TASK
1. Divide the class into groups.
2. Give out the sheets which you have prepared, or use an OHP sheet if you prefer.
3. Allocate or ask the group to choose a group leader.
4. Inform the group leader that they are responsible for ensuring that the group completes their work on time and prepares a performance for the rest of the group. The leader will need to discuss roles with group members.
5. When each group is ready, ask them to perform their playlet/mime for the other groups in the room. Each watching group is allowed to ask one question about something they did not understand from the performance. They must agree on the one question.

SUGGESTION SHEET
You are going to do a small play. The play is to show other people in your class how plants use their flowers to make new seeds and a new plant next year.
You will need to remember the names of parts of the flower, such as:

stamens
stigma
pollen
ovary
fruit
seed

The South East Essex College of Arts & Technology
n Road, Southend-on-Sea, Essex SS2 6LS
0702 220400 Fax 0702 432320 Minicom 0702 220642

> You will also need to describe pollination. Remember it can be an insect, or the wind that pollinates a flower, depending on the type of the flower.
>
> You can do the play as a mime if you wish, so that members of your group carry a card to say who they are. Or you can choose someone to be a narrator, to tell the story of what is happening.
>
> The other members of the group will have to play the parts of the flower, show how the flower is pollinated, what happens then, and then show how the fruit and seed begin to grow.
>
> You might need to look up the work you have done over the last few weeks, to remind yourself what actually happens. Before you plan what your group is going to do, talk through what happens with each other so that you are sure everyone has the same understanding.

The focus of this task is to make pupils work in a closely collaborative group to produce a piece of communication. The communication will not work unless everyone has had the opportunity to share their ideas about what they think is happening when pollination takes place and seeds develop. This involves them in explication and exploration within the group. The final 'performance' is the explanation for others in the class. The task draws on a variety of communication skills by pupils, as well as establishing some opportunities for the development of group dynamics. The leader is particularly important in this task. It does not work unless the leader can hold the group together and allow them to feel ownership of what is happening. Because they are asked to give a public performance, those who are merely being told what to do as opposed to 'owning' and understanding the ideas involved feel less likely to contribute fully. It may be that activity 3.8 might provide some prior practise on roles and leadership before the pupils do this task.

> TASK 4.5
> COMMUNICATION SKILL: EXPLAIN
> NEW ATTAINMENT TARGET 5
>
> The Key Stage 2 programme of study states: 'pupils should be aware of the way in which sound is heard and can travel through different materials'. Our aim here is to get pupils thinking about and understanding the generation of sound by using their own bodies and feelings as a mechanism for that understanding.
>
> BEFORE THE TASK
> 1. Organize pupils into groups based upon skill levels. The most appropriate skills here are probably those of playing a musical instrument, so that each group has a member who plays an instrument if that is possible.

2. Ask instrument players to bring in their instruments if possible. For everyone else, provide tissue paper and ask them to bring a comb.
3. Provide a set of sound producers of your own for the latter half of the lesson — for example, a tape recorder with a loudspeaker, a guitar, a tuning fork, a drum or anything else that shows vibration well.

DURING THE TASK
1. Tell the pupils that first of all they will be working in groups to compose a small tune. If there is a musical instrument in their group, the owner is first to play something — a few notes for the group — and then describe what sensations the instrument creates in their fingers or lips (depending on the instrument) as they play. Allow some time for the pupils to discuss this. The group then decide on a short tune, (3 or 4 bars) which they all play together, the non-instrument players making their own instrument with paper and comb. Allow them some fifteen minutes for this part of the task.
2. When they have decided on their tune, they are first asked to talk in their group about how they think the sound itself was made, based on the sensations they felt. (Questions framed in advance may help here, such as:
What sensations did you feel when playing your instrument? Do all instruments make the same kind of sensations?)
3. Now ask the groups to report back, first by playing their composed tune, then by explaining to the rest of the class how they think the sounds were made.

What this part of the lesson does is to give pupils the opportunity to explore their own understandings of the physical sensations of sounds as they produce them.
  After the explanations, use the examples pupils have given, as well as your own collection of instruments, to collect together their understandings that sound is caused by the mechanism of making something vibrate and that we can usually hear this because the vibrations are transferred to the air which then cause the inside parts of our ears to vibrate.
4. Now ask the pupils to use tuning forks or a rubber band attached to a pencil, to see if they can hear the differences that take place when sounds are made but something other than air is made to vibrate, (such as water, oil, wood, metal etc.).
5. Ask them to design a small mime that will show others what is happening to the vibrations of the tuning fork as it touches, or enters the other substance, and how the other substance in turn then vibrates. Tell them that their mime should try to show as clearly as possible what is happening.

Communicating in School Science

This can be a noisy lesson. The composition part need not take place with all groups doing the same thing at the same time. It can be phased over a period, if different groups work on different aspects of the curriculum at different stages of a day or week. The reporting back and explaining to everyone should, however take place with the whole class. Similarly, the latter part, of exploring vibrations and explaining them through mime is probably better done one group at a time, since this is the part where teachers need to encourage careful observation of what is happening, articulation of it between the members of the group, and then composition of a mime to bring out the main points. It will help pupils to develop their understandings if you remind them of the sensations of playing their paper combs and how those vibrations felt. Although the task is not an easy one for pupils, since it involves careful observation of the speed of vibrations and using touch for observation more than is commonly the case in science, it has been a very successful activity in enhancing understanding. Asking the questions about what they see and feel is a crucial party of the activity and thus makes the second part best done in groups at different times rather than a whole class task. There are clear examples of explication, exploration explanation in this task. It is the exemplification of many of the aspects of constructivism that we discussed in Chapter 1. It is not an easy task, and there are several parts to it. Never the less, the rewards lie in the skills which it encourages pupils to develop, and in the furtherance of their own scientific thinking, by focusing on the 'why and what' questions of observable phenomena.

TASKS FOR KEY STAGE 3
TASK 4.6
COMMUNICATION SKILL: DISCUSSION
NEW ATTAINMENT TARGET 4

The programme of study at Key Stage 3 requires that pupils should be given opportunities to understand the limitations of scientific evidence and the provisional nature of proof. The aim is to get pupils thinking about the explanations for phenomena and making their own explanations explicit. These are some of the most exciting and difficult bits of science. Look out for a rewarding, noisy and exciting lesson!

Clearly there can be many experiments used. The object of the exercise is to get pupils concentrating on the 'why' of science and being able to articulate their own 'why'. The experiment we have chosen here is a very simple one, focusing around kinetic theory and drawn from a piece of research from the mid 1970s. Our reasons for choosing it are that we can show the work of other pupils to demonstrate the point. Any experiment in which the group members contributed their own thinking would do a similar task.

National Curriculum Tasks in the Classroom

Look at the flask below. It contains air.
Assume you have a syringe (evacuating pump) which you can fit to the flask to take out half the air.
Pretend that you have a pair of special effects glasses which will enable you to see the air left in the flask.
Draw what you would see after you had taken half the air out.

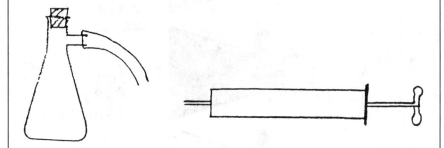

Figure 4.1: Workcard for an experiment on kinetic theory

DURING THE TASK

Give the pupils an example of the diagram above. Tell them they are to work individually at first then in groups. Divide them into the groups you want. In choosing the group members, the dynamics are the most important feature to remember. One loud pushy individual in a group can seriously disrupt the exercise. It is better to have two. Once again, the gender balance can be crucial, especially if the girls in your class are in the habit of deferring to the boys in science.

Give them these instructions for the individual work.
1. You have been given a magical pair of spectacles. These help you to see air which is normally invisible.
2. Look at the diagram. It shows a flask, containing air, and attached to a vacuum pump. The vacuum pump is switched on. It takes out half the air in the flask and is then switched off but left connected so the air cannot get back in.
3. Draw the diagram, showing what you can see with your magic spectacles after the pump has been switched off. (In other words draw where you think the air will be in the half full flask.)
4. Now get into your groups. Each one of you must show your diagram in turn to the other members of the group and explain what you think is happening in the flask after half the air has been removed, and *why it looks the way you have drawn it*.

Once the group have had the chance to discuss their ideas, give them Figure 4.2 (below). Ask them to decide as a group who they most agree with, and why, and eventually to come up with one explanation

Communicating in School Science

	1 David	2 Sara	3 Ruth	4 Gideon	5 Miriam	6 Dan
Description	Air remains on the bottom; above it there is a vacuum	Air fills the flask, but there is less of it	The air that is left is on the top; below it there is a vacuum	The air remains near the side arm	Most of the air is on the bottom, then less and less and on top — a vacuum	Air fills the flask, but there is less of it
fore evacuation / after						
Reasons	The air sinks because its specific gravity is greater than the vacuum	A gas flows, so the air flows to fill the flask	Air has nearly no weight; very light things rise	We pulled the air from this opening; the remaining air concentrates there and wants to push out	It's like what we learned about the atmosphere in our geography lessons	This is like the second drawing, but it would look like this if a little dwarf could get in and see

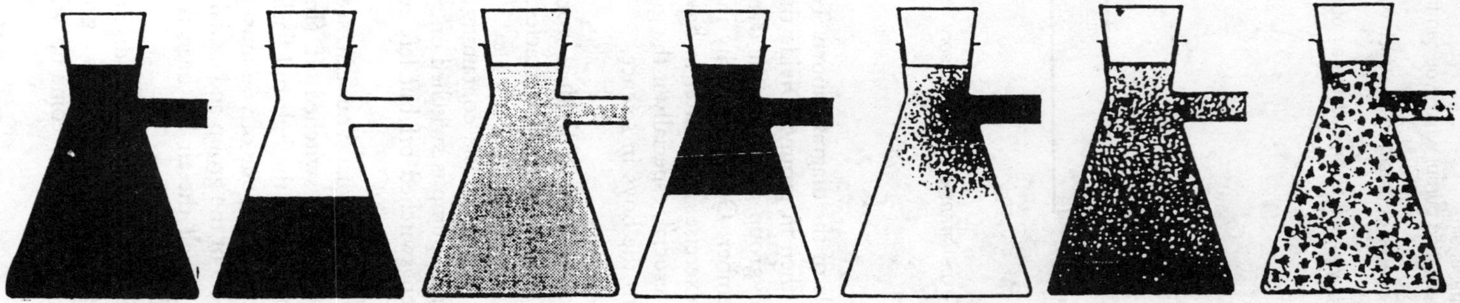

Figure 4.2: Follow-up workcard for an experiment on kinetic theory

National Curriculum Tasks in the Classroom

for what is happening for the whole group. They are to draw this, write their agreed explanation below the drawing and share their ideas with the rest of the class.

This task generates a wide variety of discussion. Some of it can be very heated. The teacher's role is to ensure that pupils obey the rules of discussion set up in advance and listen to and respect each other's viewpoints. It is crucial that rules are obeyed in this exercise, otherwise it easy for several pupils to feel that their ideas are 'wrong', especially where the group dynamics put them in that position easily anyway. Follow-up activities to this can be testing out different ideas to see if they conform with observations, convincing others that their ideas are the best explanation etc. There are a whole variety of different approaches to the follow-up of this. The issue of discussion is, in itself, worthy of a whole book. We have touched on one small aspect of it here.

TASK 4.7
COMMUNICATION SKILL: SUPPORT A VIEW
NEW ATTAINMENT TARGET 3

The programme of study here requires that pupils should evaluate early ideas about the Earth. This lesson is intended to draw together historical and scientific aspects focused around evidence for the proposition that the earth is round. It would come as part of the culmination of teaching pupils about the features of the solar system and the major characteristics of the universe. The actual carrying out of the technique would take about an hour, and the preparation a similar length of time.

BEFORE THE TASK.
1. Ensure that you have available slides/photographs of Apollo space shots, pictures of the moon, pictures of the Earth from the moon, atlases, and other types of visual evidence that pupils might want to use.
2. Prepare for each group a briefing sheet containing information such as the examples below, to help them in the preparation lesson. Several groups can tackle the same area, so you do not need many of these sheets.

EXAMPLE BRIEFING SHEET
You are sailors. You have sailed in wooden sailing ships and watched other ships. You know that when you are out at sea, other ships come into view slowly, a bit at a time appearing over the horizon. Use this knowledge to help convince the meeting that the Earth is not flat. You

will need to demonstrate your knowledge to the meeting. It is not enough just to tell them that you have seen it.

EXAMPLE BRIEFING SHEET
You are an Australian scientist. You know that when you drop something in Australia, it falls down towards the Earth in just the same way as it does in England. Use this knowledge to help convince the meeting that the Earth is not flat. You will need to demonstrate your knowledge to the meeting. It is not enough just to tell them that you have seen it. (Other ideas can be found in Driver *et al.* (1985) chapter 9, p. 170)

THE PREPARATION LESSON
1. Divide the pupils into mixed gender and ethnic groups.
2. Tell pupils that in the next lesson you are going to take them on a trip in a time machine, back to the year 1603. (A sort of 'back to the future' 4!) The time machine will deliver you into a meeting of the National Geographic Society. The meeting has been set up to discuss the argument which is currently taking place amongst sailors, the clergy and scholars on the question 'Is the Earth flat?' When they get to 1603, they will find that you are the chairperson of the Society who will have to make a judgment on the evidence presented to you. Because this is a time machine, they have the opportunity to take evidence from the present day with them, to try and persuade the meeting in helping to answer the question 'Is the Earth flat?' This lesson is to give them time to decide what evidence they will take. Each group can take a maximum of two pieces of evidence only about the area they are given. They may take some visual evidence — photographs, slides or a short piece of video material, etc., and some other kind of evidence — measurements they have made, a demonstration they can do etc.
3. Give each group a briefing sheet, prepared in advance.
4. Give them the rest of the lesson to collect the evidence and prepare their argument for the meeting. They will need to think about who plays which role in the group — leader, presenter etc.

'TIME MACHINE LESSON'
1. If you are feeling adventurous, you might want to begin this lesson with some sound or light effect which creates the impression of time travel. You might also want to wear something appropriate to the role you will be carrying out.
2. Prepare some cards in advance, indicating to the rest of the class what each group is, e.g., sailors, Australian scientists and so on.
3. Tell the class that each group will have five minutes to present their evidence. Each group can be asked three questions, one by yourself and two by any other group in the room. The questions must help the group to convince the meeting that the earth is not flat.

You will need to ask some very pertinent questions here, to convince yourself that the pupils themselves believe what they are saying. The questions may need to be of the 'what if' type in order to test out their concepts, and if there is insufficient time for the group to provide a cogent answer, it is worth getting everyone to answer these for homework, or during the next lesson.

4. At the end of the lesson, ask all the pupils to help you decide. Ask them if they are persuaded that the earth is round, and which piece of evidence proved the most persuasive for them.

Pronounce your judgment.

KEY STAGE 4 ACTIVITIES
TASK 4.8
COMMUNICATION SKILL: EVALUATE
NEW ATTAINMENT TARGET 5

Key Stage 4 requires that pupils are encouraged to develop activities which involve the critical evaluation of data.

In the task we describe below, the project is set up on a whole class basis and different groups in the class are asked to choose different materials for the construction in order to bring an evaluation to bear on their work.

PART ONE
The whole class is given the following design brief.

You have £5.00 to spend. Out of that you are required to make a toy aeroplane, powered by a rubber band. You can use any materials you wish for its construction, but it must fly in a straight line, and be suitable for use by a 7 year old. You will be required to decide at the end of your construction whether your aeroplane meets these criteria:

1   It is powered by a rubber band.
2   It flies in a straight line.
3   It is built from suitable materials which help it to do its job effectively.
4   It did not cost more than £5 to make.
5   It is attractive enough to appeal to a child of 7.
6   It is strong enough to be played with by a child of 7.

1. Once the whole class has had the design brief, divide them into groups.

2. Ask each group to decide what material they are going to use for the construction. (You need to prepare a list of the materials you have to hand: plastic, paper, balsa wood, metal etc.) The full range should be covered by the class so that different groups prepare toys from different materials. This may mean everyone reporting to the whole class what they are going to use, and some negotiation taking place.

3. Ask the pupils to return to their groups and decide on the roles in the group. Two roles are essential — leader and evaluator. The evaluators will be the ones who will act as continual judges for their own group, and a panel at the end. A briefing sheet is provided for them.

PART TWO

This may last the rest of the lesson or as long as you decide, for drawing the plans, building and testing the toy and finally painting or decorating their final version.

PART THREE

This involves trying out the final versions. All the evaluators from the groups get together as a panel. They each have a sheet like the criteria one above, and have to give a mark to each toy before it flies (criteria 1, 3, 4, 5) You may wish to get the evaluators together before this to decide on the range of marks they will award for each criterion.

2. The trial flight takes place and assessment of criterion 2 occurs.

3. The class discussion is held, with reasons for scores being debated and a focus on materials used, designs and the processes of trialling being the main features of the discussion.

BRIEFING SHEET FOR EVALUATORS

1. Your role as an evaluator is to remind the group as they do their designs of the criteria against which they have to operate. For example when they decide on what material to use, remind them that the aeroplane needs to fly and be strong enough to stand up to a 7 year old playing with it.

2. As they try designs and trial flights, you need to help them by making suggestions for improvements and telling them when you think they have succeeded.

3. You will not be the only person making suggestions, and you must try hard not to be a 'killjoy'. Focus on the positive things — for example if it doesn't fly, say 'if we made it just a bit lighter it would probably stay up' — rather than 'it's far too heavy'. The skill of the evaluator is to make the group feel positive and successful, whilst helping them to change.

4. At some point, your teacher will probably want to ask you to get together with all the other evaluators, as a group to discuss what marks you are going to award at the end and how you will decide this.

National Curriculum Tasks in the Classroom

There is a great deal in this task, and it has several extensions. The design brief can be made more difficult, for example the plane has to land in a square of a certain size. You could organize for the group to take their planes to the local primary school for the final test flight and ask 7 year olds if they would play with them — let them have a go perhaps. Or the groups could explain how they made it and why they designed it that way to groups of primary pupils.

The task clearly has a great deal of potential for working in conjunction with technology and achieving many of the attainment targets for this subject area. The costings you may have to provide for pupils if they are not actually going to purchase equipment, unless they come and 'buy' it from you. The evaluator could act as an auditor too.

TASK 4.9
COMMUNICATION SKILL: EXPLAIN
NEW ATTAINMENT TARGET 3

Key Stage 4 requires pupils to be able to distinguish between generalizations and predictive theories. This lesson should be the culmination of pupils' work on the connections between weather and physical phenomena such as air pressure, condensation, evaporation, radiation from the sun and its absorption by the atmosphere, heat of the earth's core, convection currents, specific heat capacity of the earth, gravity, jetstreams etc.

BEFORE THE TASK
1. Prepare some weather evidence — satellite photographs, newspaper cuttings showing weather maps etc.
2. You might like to take pupils on a trip round a local meteorological office or university department as part of the preparation.
3. Prepare for each group an example containing:
— a weather map /satellite picture
— a list of useful hints as to where they might look for scientific evidence to support their ideas.

Focus each example around one phenomenon on the weather map e.g. thunderstorms, snow, differential temperatures/winds at the coast compared with inland, strong winds. (Ideas can be found in Walker, 1984.)

DURING THE TASK
1. Tell the pupils the local TV station have created a new section, that of weather education team. The successful team will be responsible for presenting the weather on children's television and also explaining to children how and why the weather phenomena work scientifically. Each team will have to use their provided example of the thunderstorm etc., to prepare an explanation of why this weather occurs from a scientific

> viewpoint. Since the team are presenting science for children's television, they will need to demonstrate to the audience the science of their explanations. Divide them into teams.
> 2. Each group is to decide on the roles people will play, prepare its explanation, supported by a demonstration as part of the interviews for the posts. Each interview will last for three minutes.
> 3. You will be the interviewer.
> 4. Allow the groups time to look at the hints and discuss their explanations, descriptions and demonstrations (of air pressure, condensation, radiation etc.)
> 5. Either in this lesson or the next, provide time for them to have their 'interview'.

Summary

In this chapter we have covered a variety of examples of communication skills and student's understandings of the dynamics of groups. In the next chapter we lead on to the further development of communication skills and the growth of teams, particularly through problem-solving activities. We have tried, over the last three chapters to describe activities which will enhance the opportunities for encouraging students' communication skills and through these their understanding of scientific ideas. In doing so we have drawn upon two basic assumptions. First, that communication skills cannot be left to chance. They require direct skill teaching. Second, that to teach communication skills with any degree of success means thinking carefully about the grouping of students and the ways in which they are encouraged to work. As we explored in detail in Chapter 2, this implies a classroom organization whose primary function is the promotion of collaborative learning. As Brandes and Ginnes (1986) say:

> Dialogue itself is not enough, nor is interaction; the quality of that interaction is also crucial. (p. 15)

Organizing a classroom in which collaborative learning is the base line may mean a great deal of rethinking. Certainly it implies that planning ahead, careful and meticulous organization, and preparation is required before teacher and students enter the classroom. It also implies a great deal of responsibility, for both students and teacher. In collaborative learning situations, teachers must take responsibility for helping students to own the learning occurring, without being laissez faire or permissive. This takes courage. Students also have to learn to take responsibility; for thinking, rather than waiting for the teacher to produce the 'right answer'; for taking decisions and negotiating those with others in the whole class or small group; for owning their own progress. This may mean:

- reorganizing the physical features of the learning environment, for example, labelling cupboards clearly so that students can get what they need without asking first and giving students the responsibility for ensuring that others can use them effectively too
- agreeing with students which cupboards need to be kept locked for general safety and to come within requirements placed on teachers from outside
- negotiating the behaviour policies used by students in science laboratories
- working to agreed 'ground rules' about the way individuals interact in a classroom and ensuring that infringements can be challenged
- rethinking the organization of chairs, or in the case of fixed bench labs, small areas of the classroom where students can work undisturbed and maintain eye contact without barriers
- trying new methods of teaching designed to assist collaboration — for example, using methods of beginning and ending lessons that emphasize the need for agreement, sharing of ideas, remind students of relevant skills for the lesson.
- sharing objectives, goals and purposes of lessons with students
- sharing responsibility for learning and progress with every member of the class, so that everyone realizes that they have a responsibility for their own progress and those of others in their small group or the whole class. If one person is finding a topic difficult to understand or work at, everyone in the group needs to get together to help them — not simply the teacher or the person themselves
- thinking carefully about the use of language when talking to students. Brandes and Ginnis (op. cit.) have a good example here. They suggest (p. 54) that instead of 'ought, must, should, type statements' to students for example 'this homework should be handed in on Wednesday', students can best be encouraged to participate by being involved in negotiating deadlines
- 'lets agree about when the deadline ... will be'. Similarly, they say, teachers should not create deadlines for themselves that are unmanageable, but should negotiate these with the group too.

As far as the teacher is concerned, this means listening to students, being honest with them about what needs to be done, what progress is being achieved and how this can be enhanced. It also means developing in the classroom a negotiated meaning for collaboration and ensuring that all members of the group, whatever the size are able to separate the behaviour of a person from the essential self of that person. That is, sometimes individuals will not apply the agreed groundrules, their behaviour on that occasion will be a matter for comment by others. But the person themselves remains fully accepted and respected for themselves. Teachers and students alike need to be brave to progress to this point. But the rewards are well worth it.

Chapter 5

Communication in Open-ended Problem-solving

As the title suggests, in this chapter we look at the interplay between communication and open-ended problem-solving. When youngsters are asked to undertake tasks which have no clear definitive conclusion, we are often asking them to exercise their social, communicative and team skills to the utmost.

The chapter is in four parts. First, we briefly say something about contemporary approaches to problem-solving, specifically the current style of industrially-related, technological problem-solving. We then look at the skills and processes required as youngsters engage with such problems, building on the discussion in Chapters 1 to 4. In particular we consider problem-solving in the National Curriculum. Third, we consider the longer-term development of 'project teams' and the membership, roles and goals of this way of working. We do this through descriptions of three real-life school situations. Finally, we look at the direct skill teaching of several skills to achieve some of the goals we set ourselves.

Changes in Problem-solving

'All problems' said George Bernard Shaw 'are finally scientific problems', with a greater sense of the sweep of science than perhaps many of us share. Few young people in school would see all their problems to be scientific ones: perhaps we have some persuading to do. In one guise or another, problem-solving has a long history. Puzzles, riddles, tests and trial-by-cunning are part of folklore. In more recent days, problem-solving has become part of school science and has been explored by the Assessment of Performance Unit (APU, 1984); recommended by DES policy statements (DES, 1985); it has featured prominently in the work of the Secondary Science Curriculum Review (Stewart, 1987); and is currently being promoted through the National Curriculum (DES, 1989). It is also firmly in the focus of technologically and vocationally orientated courses, from the Certificate of Pre-Vocational Education (CPVE) and Craft, Design and Technology (CDT) and the National

Curriculum: Technology (DES, 1990), onwards to mathematics and home economics, and is advocated for primary science by the Engineering Council (1985).

Over time, the problems to be solved have ranged from brain-teasers ('make four squares from these five matchsticks'); IQ style questions ('what is the next number/ letter of the following series ...'); problems in mathematics (the 'Hanoi Tower' problem; 'tessellate this shape in 3-D'); chemistry and physics ('do problems 3 to 13 from Exercise 42 at the back of the book') to larger-scale problems ('design a child's toy that moves around and makes a whirring sound'). More recently there has been an evolution in thinking which has moved the kind of problem away from the closed, single-answer type towards an open-ended, multiple solution style.

These days, the type of problem under consideration has changed, and tackling these new problems (explored, for example, in Watts, 1991) is becoming acceptable, respectable, even mandatory classroom practice. In science this move has come about through a combination of several strands: the common commitment to practical activities and 'hands-on' science; a residual influence from 'discovery learning', and the continuing search for 'relevance' (see Stewart's (1987) discussion, for example). These come together when youngsters are asked to tackle a practical problem which has no single solution and which can be seen to use and apply their understandings in science in an everyday or industrial/ commercial context.

The Standing Conference on Schools Science and Technology (SCSST, 1985), has categorized problems as four basic types:

- those of a technological nature;
- ones with a scientific bias;
- decision-making exercises that include role-play and simulations;
- exercises that mainly involve mathematics and language.

While this is useful, the categories are very sweeping, and the distinction between science and technology is not always an easy one to hold.

Problems of Ownership

As we have noted elsewhere (Bentley and Watts, 1989; Watts, 1991) we try to be more specific about the nature of problems and follow Kahney's (1986) notion: that a person has a 'problem' when she or he has a goal which cannot be achieved directly. Jackson's (1983) summary of

$$\text{Problem} = \text{Objective} + \text{Obstacle}$$

implies just one objective and one obstacle, and that the objective can be achieved simply by removing the obstacle. However most problems are more complex and can be broken into several smaller parts, as the examples below show. Breaking them into constituent parts is often a helpful guide for allowing pupils access to finding a solution, particularly for younger pupils. Some example problems might be:

- build a bridge from paper and household junk. The bridge must span a river drawn on a baseboard. The finished bridge must support a small car and allow a model boat to pass underneath it;
- magnets. Make a group fishing game or individual maze game using magnets and household materials;
- use Artstraws and masking tape to build the tallest tower possible which can support a small Lego person (the tower should be free-standing).

These are taken from local authority materials written for primary schools (Brent, 1989), aimed at Key Stage 2. An exellent source of problems at this stage is Fisher (1987). Problems for Key Stages 3 and 4 might be:

Dimmer Switch

Many homes and theatres have dimmer switches to control the brightness of lights. Design one, with the materials available, to operate safely in a doll's house.

Paper Towels

Develop a 'fair test' for the absorbency of three different brands of paper kitchen towels.

Crude Oil

Oil production sites are frequently in remote areas and so oil needs to be transported to a refinery for processing. Crude is very viscous and this can pose great difficulties in cold conditions (like Alaska). How can it be transported safely and within budget?

Rubber Seals

From the samples of rubber provided, choose one that would be best for an oil pump, to work in extreme conditions — for example, the North Pole and the Sahara.

Car Crash

You are required to test the given materials for their suitability in car body design. The car body must be reasonably light but must be able to save the life of an occupant in the case of a crash.

Soft Fruit

When soft fruits are frozen, ice crystals are formed. The crystal size is important because if they are too large they rupture the cells and spoil the fruit. Investigate the relationship between freezing time and crystal size for different soft fruits.

Grape Sugar

Grapes are sometimes partly frozen to concentrate the sugar in the must for when they are crushed for fermentation. Investigate the relationship between the freeze temperature to obtain the highest sugar level obtained by crushing.

Ball Game

Invent a game in which a ball is used but which is very different to other ball games. There may be any number of players and the ball may be of any size and material you choose. You must also design any other implements needed (nets, bats, goals etc.).

Slow Roller

In an industrial application a hollow steel tube is required to roll down an inclined plane (45 degrees) at a constant speed. It can be given an initial push and the tube can be sealed but you must design only the minimum of constraints on the tube as it travels.

Floating Egg

Make an egg float in liquid inside a jar so that the egg is 2cm above the base of the jar. There should be no other mechanical means of support.

These problems are quite different from traditional school problems from the back of a maths or physics textbook. Little, if any, relevant information is given, there are only 'best solutions' and few 'right answers', and the methods are not provided. Sometimes there are hints and clues, but no recipes for success. Some of the list can be seen to fall into several different categories: to be relevant to a traditional science curriculum, open-ended, and also be listed as real-life problems.

However, none of the problems on the list have been generated by pupils, they have all been provided by teachers for pupils to solve. That is not to say that, once work has started, they do not become 'owned' by the student, simply that they did not start out that way. There is considerable evidence to suggest that when youngsters attempt to solve their own problems, ones that they see to be relevant to be solved, the levels of motivation are much higher than when tackling 'someone else's' problem. White (1990), for instance, describes the enthusiasm of primary school pupils as they develop their own investigation of which of their teachers' coffee mugs is the 'best', an investigation that sprang obliquely from their discussions about heat.

Before we move on, it is useful to take a brief look at the CREST (Creativity in Science and Technology) project as a national scheme to promote problem-solving.

CREST

The CREST acronym stands for CREativity in Science and Technology and heralds a scheme sponsored jointly by the British Association for the Advancement of Science (BA) and the Standing Conference on Schools Science and Technology (SCSST). It is coordinated at a local level through Science and Technology Regional Organizations (SATROs) and by some local education authorities. The entire project is basically an award scheme for rewarding youngsters' efforts in school; the project's primary aim is to promote scientific and technological problem-solving in the 11 to 18 age range. In this sense, it builds on the Young Investigator's scheme, also sponsored by the BA, and which is targeted at the junior school age range, 8 to 12. Both schemes aim to complement normal school work and are non-competitive: youngsters gain recognition for their work through the scheme as a national project, and receive a Bronze, Silver or Gold award. A major point of the scheme is that youngsters work on their *own* project. They may work individually or in groups, but the project is one that they choose. Teachers may be influential at the point of choice — as may the CREST local organizer — but the emphasis is heavily weighted towards the youngster designing his or her own investigation. This is particularly so at the level of Silver and Gold where the scheme is looking for quite original (creative) work. Some of the problems in the list above are examples of CREST projects.

One of CREST's secondary aims is to promote closer working relationships between schools and engineering, industry and commerce in the outside world. Some projects have involved the electronics industry where, for example, youngsters have designed a '7-day pet-cat feeding system' so that the cat's owners can leave the pet well fed while they are on holiday; or a system for allowing only the owner's cat through the cat flap.

CREST has close associations with other 'problem-solving agencies', in particular the World Wide Fund for Nature's Environmental Enterprise Award (WWF, 1989), and Surrey SATRO (1990) who themselves have an enviable record in promoting approaches to problem-solving. Both of these agencies have developed 'science and technology' problem-solving schemes where, in WWF's case, youngsters direct their attention at technological solutions to problems in the environment.

Why Teach Problem-solving?

There are a variety of answers to this question. In the NCC (1990) we see problem-solving as an advanced and sophisticated method for developing and using a variety of scientific and communicative skills. For example:

- problem-solving enables youngsters to take ownership of a task
- it is a form of both active learning and discovery learning
- it is a vehicle for teaching many scientific skills, and for teaching the content of science
- it allows cross-curricular activity
- it provides relevance and real-life contexts
- problem-solving and creative thinking are amongst the highest and most complex forms of human action

All by itself, then, problem-solving is an important component of learning. Gagne (1970) for instance says:

> The results of using rules in problem-solving are not confined to achieving a goal, satisfying as that may be for the thinker. When problem solution is achieved, something is also learned, in the sense that the individuals' capability is more or less permanently changed. What emerges from problem-solving is a higher order rule, which becomes part of the individual's repertory. The class of situation, when encountered again may be responded to with greater facility by means of recall and is no longer looked on as a problem. Problem-solving, then, must definitely be considered a form of learning.

However, it must be learning in a context. Hadfield (1987) makes the point that in problem-orientated teaching the teacher structures the material in such a way that problem-solving situations arise as a natural part of classwork. They are an integral part of the on-going exercise in science. Problems in one context can then be seen as appropriate for use in another so that, as Gilbert (1987) says of her own work with young children:

> problem-solving exercises must be directly and obviously applicable to any previous discoveries, or the children may be confused by the change of direction, discard the previous information as useless and give up any attempt to reason for themselves.

In summary, then, problems can be used to stimulate work

- at different times: the beginning, middle or end of a topic of work
- at each key stage — there is no age limit
- as a way of tackling New Attainment Target 1 in Science
- which builds on individuals' prior knowledge and skills
- that requires the negotiation of individuals' ideas, thought, and 'world knowledge'.

Nor need problem-solving be exclusive to science and technology, of course. For example, the Association of Teachers of Mathematics (ATM, 1986) suggests that problem-solving requires pupils to have the opportunity to

- engage in discussion
- work with others in a stimulating and motivating environment
- cooperate and share ideas with others
- work as a member of a team
- engage in activities that are relevant to them as individuals
- participate in decision making
- take responsibility and act with initiative when appropriate

For us, it is this last note which provides one of the central virtues of problem-solving: it is a means of transferring some of the responsibility for learning to the learners. The main point of adopting the approach in schools is putting the emphasis on the learner using a planned approach (their own or someone else's) to tackle a problem. It becomes their responsibility to delineate the problem, decide on what an appropriate solution might be, derive and test possible solutions, and choose the point at which they think the problem has been solved. As above, the problem might be one that is presented to the solver or one generated by their own thinking, actions or lifestyle. If youngsters are not engaged in setting the goal and working out the routes and strategies for themselves, they may not see the goal as worth achieving. Increasingly, problem-solving is seen as a valuable way of providing more open learning situations, where youngsters are less constrained by didactic teaching methods (Bentley and Watts, 1989). In our experience, problem-solving is an excellent motivator and it breeds enthusiasm — even in some of the most traditionally de-motivated youngsters. We have other reasons, too.

Communication in Open-ended Problem-solving

Problem-solving can be undertaken by individuals and by groups: it is the latter that interests us here. Ideally, problem-solving encourages the development of a team approach to solutions, thus it serves as a perfect vehicle for a variety of communicative acts and (thus) the skills of communication. In this sense, then, it:

- enables team groupwork and many social skills
- encourages decision making
- enhances communication.

Problem-solving and the National Curriculum

A key requirement of the National Curriculum is that pupils should be encouraged to develop their investigative skills and their understanding of science through systematic experimentation and investigations which (by age 13) are to be:

> set within the everyday experience of pupils and in wider contexts, and which require the deployment of their investigative skills and the use and development of scientific knowledge,

and, by age 16, are to be:

> set in the everyday experience of pupils and in novel contexts, involving increasingly abstract concepts and the application of and extension of scientific knowledge, understanding and skills, where pupils need to make decisions about the degree of precision and safe working required.

The new version of the National Curriculum (NCC, 1991) goes on to require amongst other things

- pupils to plan and carry through investigations in which they may have to identify, describe and vary more than one key variable
- require pupils to make strategic decisions about the number, range and accuracy of measurements, and select and use appropriate apparatus and instruments,

and that science work should:

- promote invention and creativity

In the main we prefer the more explicitly 'communicative' version of the 1989 document where, for example, level 6 pupils were asked to:

contribute to the analysis and investigation of a collaborative exercise in which outcomes are derived from the results of a number of different lines of inquiry, possibly including experimentation, survey and use of secondary sources, in the context of which each pupil should:

- use experience and knowledge to make predictions in new contexts;
- identify and manipulate two discrete independent variables and control other variables;
- prepare a detailed written plan, where the key variables are named and details of the experimental procedure are given;
- record data in tables and translate it into appropriate graphical forms;
- produce reports which include a critical evaluation of certain features of the experiment, such as reliability, validity of measurements and experimental design.

While much of this clear emphasis on communication and problem solving is lost from the latest document, both skills re-appear in the National Curriculum Council's (NCC, 1990) Crosscurricular themes and dimensions. They are presented as crosscurricular skills which are 'absolutely essential' to be 'located in schemes of work for each key stage and supported by an overt policy on the teaching methods required.' Here, then, are some of our suggestions.

The Skills and Processes of Group Problem-solving

First, in a broad sense and expressed as a series of 'can do' statements, these are:

- can find and use information from a wide variety of sources, some given, some sought;
- can find patterns of information in given data, and can decide upon relevant information from a variety of sources;
- can use software to log data, represent situations, form and test hypotheses;
- can make and adapt plans to meet changing circumstances, can choose an optimal way of tackling a task given alternatives;
- can understand and follow instructions, and can make suggestions to adapt procedures;
- can perform complex practical tasks with reasonable accuracy, precision, dexterity and coordination;
- will volunteer information in a clear and organized way. Can describe personal opinions, observations and arguments adapting language to context and audience;
- can listen sympathetically to others and show understanding of diffe-

rent arguments and viewpoints. Is sensitive to non-verbal cues, and can evaluate the ideas, opinions and attitudes of others;
- can write in the style and vocabulary suitable to the task, and can create deliberate effects through prose;
- can show imagination and creativity in producing ideas, and can display originality, richness and flair;
- can describe space/form in 2D and 3D, and has an appreciation of perspective;
- can use straightforward and specialist instruments for measurements, with an appreciation of scale, dimensions and errors;
- has knowledge of, and can use, basic tools and materials carefully and safely;
- can approximate and estimate quantities to an accuracy suited to the task in hand;
- is able to make decisions and participate in group decision-making processes, can cooperate and sustain the decisions of the group and the group effort;
- can respond to the group's need for leadership with sensitivity, with acceptance from the group, and work with other leaders;
- is able to help others extend their self-awareness and personal effectiveness through self-knowledge.

These are quite general skills so that, for example, we have not specified which measuring instruments, tools or materials so we can imply a wide range of circumstances. Moreover, these skills are fairly idealized in that — should we stumble on someone who is proficient in all these areas — s(he) would have competence indeed! Nevertheless they are component skills to problem-solving and we must look to teach these within science. Moreover, it is possible to see some of these competencies displayed in good classwork, and we highlight some of them in the two 'situations' we outline below.

Nor are these skills serial, so that they do not have to appear or be realized in this — or any other — order. Different parts of the problem-solving process will draw on different skills at different times.

Groupwork and Teams in Problem-solving

Here, as before, we draw a distinction between a team and a group. A group is one which may or may not have a long time working together on a task. Frequently, the group lasts only for a short time — for the duration of a lesson for example — or for a very short time to undertake a specific learning activity, say, conducting part of a discussion. A team however is quite different, it is a group which will be in existence for a long time (weeks, months). The team needs to grow and develop to accomplish its task effectively. It needs to invest time in getting to know the strengths and weaknesses of team members, learn how to work together, be honest with each other and support one another. Often pupils are part of a team at home so they can have some experience of such situations, although, clearly, a family is a very special kind of team and parallels with home experiences are not always appropriate.

Looking at what is involved in team problem-solving, we return to the main summary of the framework we established in Chapters 3 and 4:

- groups are constructed for both general and quite specific purposes
- teachers construct groups with short-, middle- or long-term duration in mind
- the composition of the group determines the outcomes for which they have been constructed
- groups can be constructed with a variety of membership criteria and norms in mind, and can be homogeneous or heterogeneous in respect of those criteria and norms
- groups need to be taught how to operate successfully as a group
- the development of self direction, autonomy, group ethos, role-play and some norms are best managed through long-term group stability
- we reserve the term 'team' for a group constructed for particular purposes but with long-term outcomes.

Constructing a Team

The construction of a team is:

- based on the teacher's knowledge of the class
- based on specific purposes in mind
- the class's familiarity with being placed in groups — as a common or infrequent occurrence.

ACTIVITY 5.1
(Suitable for Key Stage 3 pupils)

Imagine your science class is similar to classes over the country. You will have some youngsters keen to be involved and, at the other extreme, a rump of characters who will cussedly resist being motivated. In between will be a cohort of pupils who will take up the task but need encouragement to feel involved. Let us imagine, too, that the class is in the throes of a topic on mechanics and forces and you set the problem:

> Design and make a portable device for old and infirm people to lift a milk bottle from doorstep to worktop level — without the need for them to bend over, or to use electricity (borrowed from Surrey SATRO, 1990).

BEFORE THE SESSION
1. Make some prior decisions about team members. Chapters 3 and 4 give a variety of hints and suggestions about membership and the added ingredient here is that you would expect the group that you are constructing to develop — grow into a team.
2. You may also need to draw up some 'role sheets' and instructions for team leaders. If you have decided who the team leaders are to be, it may help to bring them together in advance, or — as part of their homework — give them a 'team leader's pack' to study, with hints in it about group processes to help them with the roles they will have to assume.

THE FIRST SESSION
The class needs to know
- what the task is
- who the teams are
- what roles the task dictates
- how it will be assessed
- what the deadlines are
- the short-term and long-term periods of time available to them.

FOR THE PUPILS
1. Hand out the task.
2. Divide the pupils into teams
3. Ask each team to brainstorm keywords on a piece of paper about what is needed for the overall task to be completed. (They may need to practise 'brainstorming' before the event)
4. List different ways they think the task could be done, all together, or on a large sheet of paper to show others in the room;
5. Split into pairs or threes within each team to list what resources they will need for the task, and then pool these ideas with the others in the team;
6. List what needs to be done in the short term (by next lesson), and shape these so that there are as many tasks as there are people in the team (without cutting vital tasks and adding spurious work). The team must now decide who is to take each of the preparatory tasks and the time scale for completion. If someone in the team lacks the clarity of direction, or the resolve, the team must provide ideas for them to move forward.
7. Sort roles on a short or long term basis
- group leader
- recorder
- reporter
- technical organizer

These can stay the same or rotate. Only one group leader role, but there can be multiples of the others.

8. Decide who and how the first task assessment will happen (self, peer, group or teacher assessment).

FOR THE TEACHER

1. The key task for teacher is to get around the teams to help sort differences of opinion, egg-on slow workers, keep the class to time, guide resource implications, channel some of the enthusiasm and energy and — above all — to praise the efforts of each team.
2. Use the opportunity to spot potential team problems developing early on, and if possible nip them in the bud by advice or moving team members. The choice of team members will seem ideal in some cases and the team will set off with gusto. The 'chemistry' of other teams may seem less ideal and now is the first chance to see what is happening.
3. One area that might be problematic is number 5: the roles to be played. One way is to give them a 'role sheet' so that they can see what is involved. These can be changed over time, but form the basis by which they work. Some examples are given in Chapter 2 (more below).

SECOND SESSION

1. report back in teams to each other on what has been achieved (reporters' role). They direct that part of the activity:
2. sort out the team rules (take turns to scribe, take turns to share out the jobs, shared discussion, ways of working);
3. assess individuals' performance so far;
4. plan the next moves; and
5. set to work;
6. assess progress at end of session and plan for the next session.

The third of these is likely to present most difficulties to the group unused to peer or self-assessment. We explore some of the aspects of assessment in more detail in Chapter 6.

Managing Problem-solving

Managing problem-solving in the classroom this way raises a multitude of questions. For instance:

> At what point do we teach pupils about group dynamics? How do they sort out the problems — how can they focus on what is happening inside the team as opposed to how well the task is progressing? How can they sort out the argumentative ones, the trouble makers, the late producers, the over eager etc.

The sort of sheet above is an opportunity to focus on the dynamics rather than the skills that people in the group possess (or do not possess). A possibility is for one member (in turn) to step out of role and take the chance to observe the group performance. Discussion is easiest to focus on. It may be best for the team to have worked together for a while before spending time looking at group performance.

> Does one set the same problem for a whole class group so that they generate different solutions? Is it best to set one large problem for all and ask different groups to tackle different aspects so that they build up a composite picture at the end? Is it preferable to set different problems for different groups, or even a different problem for each individual and allow them to work separately?

There are no formulas here and no doubt each class teacher will try different formats at different times and decide on how well groups respond to different approaches.

> Before beginning problem-solving, does one first deliberately teach all the facts, concepts and skills so that the pupils will have all the relevant information at their fingertips?
> Or does the teacher use the motivating power of problem-solving as a means of allowing the pupils to decide and satisfy their own knowledge needs?

Again, there is an element of preference here. Clearly, if an exercise requires the use of a sensitive balance, a microscope or newtonmeter, the teacher may decide to suspend activities while everyone explores its use — or, alternatively, set up 'skill stations' so that each group can take 'time out' separately when they need it. In order to help exemplify some of the possible solutions to these managerial issues we explore three real-life situations.

Problem-solving in Action

To recap, our position in this book is that open-ended problem-solving requires high-order skills of communication and advanced skills in groupwork. In this sense, then, this chapter builds closely on aspects discussed in other chapters. Our main point is that teachers need to plan for:

- group membership
- team development
- problem-solving skills and processes
- the assessment of groupwork, and the
- management of long-term projects

- how to achieve effective groupwork to achieve particular outcomes
- how to promote communication in an appropriate way

We take three different but real-life situations (where the names and details are only vaguely camouflaged), describe the organization and setting, and examine the problems and solutions arrived at by the pupils.

Situation 1: Key Stage 2

The first school is a large primary school in the Greater London area, where the Junior staff have some experience of problem-solving, having entered pupils' work for national 'environmental problem solving' competitions. The school has quite extensive grounds with a 'wildlife area' which habours a pond and ducks. This area is largely uncultivated, fenced from the playground and is used for nature study work. There is also a small area of garden maintained by a combination of pupils, teachers, parents and the school keeper. For Class 4 (10–11 year olds, Key Stage 2, class size 30) the first stage was to decide on the problem. They began by individually writing a 'moan list', based on an idea by Adams (1974). They noted things that bothered them in their everyday lives, bothered teachers, bothered parents and adults and which in turn were often visited back on them. Their lists included 'having to brush my teeth', 'putting the top back on the sauce bottle', 'turning off lights', 'pulling up my socks', 'having to wipe my feet', 'sour milk' and so on. The point is that things which bother individuals probably bother others too, and some solution to the problem could have a reasonably wide appeal. The separate lists were amalgamated to one large sheet — while the class asked others to answer questions of clarification about their moans — two 'scribes' compiled a full list. In discussion, the teacher pruned the problems to those with an 'environmental bent'. In paring the list and allowing free discussion the class arrived at two main ideas:

- to build cloches for the garden to protect plants and seedlings from pets and pests, and
- to build an island in the pond for the ducks to nest in peace.

These did not surface directly from the 'moan list' but grew from airing a wide variety of problems, the list was just a stimulus. The teacher split the class about 50:50 to consider both problems. They were (roughly) free to choose which problem they would tackle and then organized into temporary groups of four and five. The processes were very similar in each case: they initially brainstormed around the problem, sorting what question needed to be answered and tasks needed to be done. Which were the pests that were attacking the garden? Could one cloche act to resist all known pests from pigeons and rabbits, to slugs and snails? What would they be made from?

How could they be tested? What were the commercial types like and how expensive? What size would the island need to be? Would it be permanent or a floating platform? How could it be anchored? What would need to be on the island?

The groups of four now came together within each problem and shared their ideas. Again, two (different) pupils 'scribed' and compiled the full list of questions and suggestions, to be read out to everyone. The time now came for a re-shuffle: individuals had to choose which problem they wanted to do. They 'signed up' at the end of the lesson, most opting for the problem they had already began work on; some moved back and forth before settling for one. Before the next session, the teacher designed six (long-term) teams of five and produced brief role cards for each team. These were:

- Director — chair of discussions, time-keeper, arbitrator
- Go-getter — responsible for acquiring materials, apparatus, tools, equipment
- Site manager — responsible for tidiness and safety, arranging the out-doors work, traffic warden around the school
- Recorder — maker of notes and lists, writer of letters, recorder of measurements
- Accountant — keeping a tally of costs and materials used.

The class had one morning a week devoted to the work although it grew to dominate lunch, break and evenings too. The teams rotated the roles each week and shared out the other jobs as they arose, the Director making final decisions in each point of dispute. Individuals kept their own folders of the work they did and generally relied on the Recorder for a running summary of what had taken place.

Over the next half term both problems were undertaken in a wash of enthusiasm and activity. Letters were written to manufacturers; scale drawings prepared; materials acquired; models tested to destruction; parental help came in the shape of 'muscle power', tools and expertise; prototypes were produced and trials accomplished; accounts were written; measurements recorded; photographs taken, and display work mounted. Of the two, the Island Problem was by far the more difficult to realize and the outcome least satisfying, the moorings of the final model broke in high winds and the damage was sufficiently disheartening to eclipse further serious work.

Each phase of the problem could be described in detail but here we focus on one: the design stage for the Island Problem. Each member of the group was asked to draw their impression of what the island would look like and to number the features they had introduced. The most obvious difference was in a permanent island and the 'raft' type. Here the teacher organized two (temporary, short-lived) groupings, mixed in allegiance between 'permanents' and 'rafts', to explore the cost-benefits of both. Each sub group had to examine one of the two methods and they began by listing its main properties and setting out some of the difficulties of putting these into practice. They had to propose alternatives, explore these too and reach a final decision on one of the two approaches. The final decision-making process came as a forum, as both sub groups related their discussions and made their decisions known. There proved no need for a vote because both ruled out the permanent island as too difficult to do; the raft more interesting to make. More brainstorming followed, on the size, shape, materials, mooring, fixtures. Then more individual designs and different sub groups organized to cost-benefit the ideas until the final design was agreed. Now there were three teams to tackle the raft, its mooring, and its fixtures and furniture. The roles in each team were team leader, 'materials getters', 'accountants' (to watch over amounts and costs), 'recorders', and each of these rotated with time. At all times everyone was a 'doer' and a 'tester'. In this phase, each team had to generate three different but viable alternatives, giving a possible twenty-seven final versions. Not all twenty-seven were tried; each team took a version and constructed a model for testing. This led to agreement as to the final shape of things and construction began in earnest. The three teams remained intact and focused on their own part of the problem. There were frequent 'reporting' sessions when — with or without teacher — they discussed progress and shared ideas and help. All the while, of course, the 'cloches' half of the class were undergoing similar processes and there were whole-class summaries and evaluations, too, of each others' problem. Roles continued to be rotated so that some children would be the 'day-leader', take turns using the camera, find ways to acquire materials et cetera.

Situation 2: Problem-solving at Key Stage 3

This took place at a large comprehensive school in Surrey, where classes are reasonably familiar with problem-solving techniques, particularly in their technology work. In this case, it was a class of twenty-seven, Year 7, co-educational, mixed ability pupils and from differing backgrounds. The school has entered pupils for local 'egg-race' challenges with success (see Watts, 1991), and their science course calls for problem-solving on a fairly regular basis. Here are three problems used, addressed to the pupils:

Problem 1: The 'Hoverloon'

Brenda and Po Chui had just come back from a day trip to Calais on a hovercraft. Unlike the rest of their fourth-year French group, they are arguing about the way the vehicle is able to both hover and travel. Brenda bet she could make a balloon-powered hovercraft which would beat anything that Po Chui could make. They agreed they would only use materials they could

Communication in Open-ended Problem-solving

get from school and that the craft must be supported, and driven, by the air from a balloon and that they are not allowed to push it at the start.

1. What practical problems do you have to sort out first?
2. What would be a fair test of the best Hover-loon?
3. What apparatus and materials will you need?
4. Draw your plans before you start and make notes about any changes you try as you do them — you won't remember later!
5. If you had more time or could start again, what would you do in order to make your Hover-loon go better, further?

Problem 2: The Two-times Problem

Sharmi and Ceris are talking about clocks. They want to work together to make a timing device that can do two sets of timings, it can accurately signal when twenty seconds has elapsed and stop when a minute is up. They can only use the materials they can find in the classroom, although they have found a wooden stand with three holes in it How would you make the clock?

1. What practical problems do you have to sort out first?
2. What would be a fair test of the timer?
3. What apparatus and materials will you need?
4. Draw your plans before you start and make notes about any changes you try as you do them — you won't remember later!
5. If you had longer to work or could start again, what would you do in order to make your timer more accurate, more reliable?

Problem 3: The Marble Method

Eddie and Carmelene have spent the day with a small local company which makes toys and games. One game needs a thirty second delay device. The production team say that it has to be started by a marble and that the end of the thirty seconds is signalled by a marble being ejected from the toy. The team says there is further money available for development if they get a good idea. Eddie and Carmelene have only the materials they can find around them, plus a few marbles from home. How would you make a delay device?

1. What practical problems do you have to sort out first?
2. What would be a fair test of the delay device?
3. What apparatus and materials will you need?
4. Draw your plans before you start and make notes about any changes you try as you do them — you won't remember later!
5. If you had longer to work or could start again, what would you do in

order to make your device more accurate, more reliable? Could you make one that produced a much longer delay?

Three problems, three groups of three pupils each to tackle each one. Four double lessons are given over to the work at the end of the Autumn term, just before Christmas. The first is given over to discussion and planning and in the next three the groups will tackle each one of the problems in a 'circus' cycle. The teacher decides the teams, who will stay together for the four sessions. The roles are:

- Team Organizer (leader, timekeeper, 'final say-er')
- Team Arranger (materials, tools, apparatus)
- Team Auditor (scribe, monitor, reporter)

Like the problems, the roles rotate each week: Auditor becomes Organizer, Organizer becomes Arranger and so on. In lesson 1 the groups are announced and the roles are discussed. In this instance, all the Organizers, Arrangers and Auditors for the next lesson are brought together and — for five minutes — the nine together list and discuss the duties attached to each role. They then move back to their teams to plan briefly how they will tackle each of the problems, and draw up their lists of materials: the bulk of which would normally be found in the 'Bits Box', an old tea-chest of yoghurt pots, rolls, egg-boxes and card which is topped up on a regular basis. The questions attached to each problem help to shape the planning session.

Although they occupy the same classroom space, teams are commonly so engrossed in their own problem they do not really appreciate the trials and tribulations of the other problem-solvers around them. Each lesson ends with the teams who have tackled the same problem coming together for five minutes for the Auditors to sum the point they have reached. Homework each time is for accounts to be written of the day's work. At the end of the fourth session, the final problem-solving lesson, the teacher summarises the class's work, and each team is asked to discuss the various roles: which they found easiest, which was the most difficult, how well the team responded to the others' roles.

Situation 3: Key Stage 4

Now for a different scenario: High Cross is an 11–16 comprehensive school in a rural setting. Each year it participates in a 'Problem-Solving Summer School' organized in conjunction with the local SATRO at the nearby Sixth Form Centre. The Summer School takes place in July — in the post-exam period — and participating schools are invited to send a group of six 4th years (14–15 year olds, Key Stage 4), plus teacher. Each school group is provided with a problem linked to a local industrial concern and the industrialists are

Communication in Open-ended Problem-solving

brought in during the week to advise and evaluate. The schools are sent the problem and contact names in advance and given some indication of what facilities will be available for attacking the problem. The problems range from:

- Investigate the polyunsaturated/ monounsaturated/ saturated fatty acid content in some common foods (in conjunction with a dairy food company).
- Design an automatic brake to slow a runaway wheelchair, pram or pushchair gently (with the adjacent hospital).
- Investigate the effects of readily available household chemicals on water daphnia (with a nearby chemical works).
- Design an automatic greenhouse watering system (with a garden centre).

There are many more and in each case the schools spent time with the link-company, examining the problem in a real situation. Here we focus on some of the skills and processes of just one task:

Evaluate the performance of a commercial household 'odour remover' and design improvements to cope with an industrial context.

The problem was constructed in collaboration with a pet-food factory, and anyone with local knowledge of such a process will appreciate the problem. The company wanted an effective way of deodorizing their offices and reception area, presumably so as not to deter staff and potential customers. While one might want to question the implied values of continuing to sully the environment while designing a 'clean conditions' area for themselves, the problem was certainly live with relevance and discussion points.

When the problem was received at High Cross, the school was preparing for its own Activity Week when the timetable is suspended and a wide variety of activities are offered throughout the school. The group were volunteers from different classes (four boys and two girls) who would stay together for the full five days. They would expect to have to report on their project at the end of the week to the rest of the Summer School, and again on return to their own school at a Year Assembly. Clearly the problem was not of their choosing but they accepted the challenge of developing ownership and devising solutions.

The group met twice before the Summer School began, in order to discuss the problem and to 'talk tactics'. At this stage two teachers were involved (their usual science teacher and a student on teaching practice) who took turns to accompany the group each day. At this early stage both teachers merely helped convene the group and offered moments of advice — the bulk of the discussion was the youngsters' own. They decided on a rotating leadership role — a different person each day: the sixth would make the presentation at the end of the week — they would all share the presentation on return to school. The leader was to liaise with teachers; act as spokesperson on site, and nag the others into continued action. Everyone kept their own records and they began the rota in alphabetical order.

The problem was broken down into two consecutive parts, the evaluation of the device itself, and its re-design for a different context. The first task was subdivided into investigations of 'standard air flow', 'variation of smells', 'filter types' and 'detection methods', and the group worked separately since they decided that ones and twos could tackle the various parts independently. The 'air flow' rig was a hairdryer and hood attachment set on 'cold' and connected to a variable power supply. The filter types included the original charcoal filter from the commercial device, the insoles from 'trainer' shoes advertised as 'odour-eaters', and an assortment of ad-hoc filters made of cotton wool and other materials. The smells were taken from perfume sprays, ammonia, 'bad egg' drops from a joke shop and other aerosols like furniture polish. The detectors were the most difficult and, after a number of chemical tests and blind alleys the group decided the human nose was the only realistic solution. This, however, quickly became the task of one boy whose facial features were thought to give him the advantage of volume. His protests were dismissed on grounds of standardization (and a somewhat spurious appeal to the high status of 'sniffers' in perfume and wine industries) and, with some sheepish reluctance, he became The Nose — a role which required ritualistic walks outdoors 'to cleanse the palate' and 're-freshen the nose-buds' (sic). The two girls tended to work together except when one was the day leader; it fell to other girl to be end-of-week presenter-narrator. The team quickly established camaraderie through humour and much fun was had as they tested the smells, while The Nose was prominent and much in evidence, the effect of some odours were clearly not lost on the rest of the group.

The re-designed model comprised an 'expelaire' fan, reversed to blow air into the room, with the adapted filter in place in the metal duct tubing leading from outside to the fan. This was all assembled into a wooden board to fit flush over an open window in the firm's office. The office staff were asked to judge the effect with and without the fan in place and — generously — judged it a success. As a mark of their humour the group performed the final presentation with all six wearing false noses ranging from the comic to the ridiculous.

The development of pupils from the year 6 ones we saw in situation 1 to these sophisticated young people in year 10 is obvious. There is a great deal of complex communication happening by this stage. Not only are they capable of sustaining a problem-solving activity for a week, they can negotiate with adults, make decisions for the way a group works, report to a wide ranging audience and most importantly in our terms, develop a team approach that is capable of involving humour to assist the growth of camaraderie, and be confident enough in the support the team offers to present their humour openly to others (wearing funny noses when reporting back). In this chapter, we propose to concentrate on the development of the team approach. As we stated earlier, there are exercises and skill practice which can enhance the development of a group into a team and thereby make problem-solving a rewarding scientific experience for young people.

More Direct Skill Teaching

We now use these situations to draw out some of the potential which problem-solving, unlike any other method of teaching, has for increasing motivation and communication. In the drawing up of solutions to the problems pupils engaged in a variety of communicative acts. They wrote letters, made drawings — both in exploration of their own ideas and in explanation to others — they wrote scientific descriptions, took photographs and created displays, they wrote classwork reports, homework accounts, apparatus lists, and answered written questions. In terms of verbal communication, they brainstormed, liaised with teachers and technical staff, gave directions, issued edicts, chaired discussions, reported back to small groups and the whole class and used a democratic forum for debate to lead to decisions. In order to accomplish many of these communicative acts, they had to work as teams and share and develop a variety of roles. A fairly impressive list by anyone's standards! It is important to note that in Situation 1, the pupils were coming to the end of Key Stage 2. Thus one could expect by this point they would have acquired certain skills in, for example, writing letters, and creating displays. However, they could only have done so by means of careful progression throughout their years of primary education.

Reporting

From their entry into school, most pupils are (and should be) encouraged to report. There are very few reception and Year 1 classes where pupils do not at some point in the day gather round the teacher for the purpose of talking and listening, sharing their experiences from home for example about a particular topic which the teacher has chosen. That close supportive 'family' base is the early beginnings of reporting. In such cases do children learn to share thoughts and ideas with others, learn to listen, and to follow on from others' suggestions? At this stage of course, much of the sharing is concerned with pupils' individual experiences, they are less involved with sharing experiences on behalf of others in the class. Given this, we must deliberately foster the work of encouraging pupils to think about the listening and reporting skills they can use in small groups. It is also at this stage where the early skills of turntaking and encouraging quiet members, can be practised by the teacher. Of course many pupils remain unaware of these early skills and there comes a time at which this awareness must begin to be raised.

Communicating in School Science

In Years 1 and 2, pupils are used to sitting in groups and working on similar pieces of work. As pupils develop, the emphasis begins to move from working alongside others in a group to collaborating with them on a joint piece of work. Precisely when this change in emphasis takes place will be a matter for teachers to decide, based upon the individuals in their class. There is no doubt that by the time pupils are into Year 4 (8 year-olds) they should be collaborating and discussing joint tasks with their peers. We would suggest a progression for Years 1 to 6 such as:

- Year 1: encouragement of listening and talking about scientific experiences as individuals, being placed in small groups and a listening adult to assist whenever possible.
- Year 2: practice of listening skills through games (see Chapter 4) and involvement of working in pairs and threes through games which require cooperation. In mixed age groups, encouraging the notion of partnership with an older pupil who helps, listens and works with younger ones.
- Year 3: establishment of partnership as a common working method — telling your partner news every day — reading to your partner, working together as a pair in games with other partners. Telling the rest of the class what they as a pair have done, in carrying out an experiment for example. Movement should occur for all pupils during this year towards using groups for collaborative working. By this stage, pupils should be able to fulfil different tasks, to bring an experiment to completion for example. Even if they are working as an individual, they should be sharing a common focus which is agreed with others. In terms of reporting to the class, one member of the group should be encouraged to take responsibility for saying what the whole group did.
- Year 4: progress in listening skills through focused exercises and introduction of the notion of roles — perhaps task associated at this stage — in a group. Work should include practice of different roles and introduction of the notion of leadership. More development of reporting on behalf of the group, allowing time to compose a group report before one person delivers it to the class.
- Years 5 and 6. Further development of roles and the introduction of teams ought to be possible here. Pupils could learn about group dynamics and practise some of the implications of this for their own team work. By the time Key Stage 2 ends, many pupils should be capable, as they are now, of working in collaboration with others to bring a problem to a satisfactory and agreed solution, with an awareness of the need to work as a team.

Brainstorming

Brainstorming, as a game, can be developed quite early in the whole class group. It is dependent upon the skills of a recorder (usually the teacher in early stages) in writing down the ideas, with some notion of time. Learning the skills of listing as many ideas from as many people as possible at the start of any activity is good training for pupils as individuals. All teachers practise forms of brainstorming, anyway, when they ask their classes for ideas and suggestions which they write on the board. The movement from this to a time constrained (or member constrained: taking one idea from everybody even if the idea has been suggested before) is but a small jump. As soon as the recording skills of pupils are able to cope, brainstorming as a method of exploration and explication in a small group can begin. It is an invaluable tool: by the time pupils reach Year 5, brainstorming should be sufficiently developed as a skill that pupils decide to use it themselves without having to be told, in much the same way as they would collect and use a ruler or rubber without asking the teacher's permission.

Graphicacy

Displays, scale drawing and modelling, all involve communication skills of a visual nature. Displays are a feature of the learning environment with which pupils are very familiar in primary schools. They are very used to having their work displayed (often very imaginatively) by the teacher as a symptom of recognition and praise. As they progress through the school, many pupils are practised in helping design displays too. This provides valuable experience for making decisions about visual presentation of ideas. However, all pupils need to be involved in the decision making generally in the classroom if they are to appreciate some of the issues involved with display as a means of communication. With Year 1 pupils, this might be something as simple as asking them what colour background would be best for some work. As pupils get older (say Years 4 or 5) it might mean handing over the responsibility for a part of the classroom display area to different groups of pupils who then have to display their own and others work according to criteria such as:

- the display must have a theme connected to aspects of the work being carried out in the classroom;
- everyone has to have a piece of work displayed at some point during the half term;
- the display has to be aesthetic, eye-catching and well presented;
- there has to be a mixture of pupils' work and other displays which are concerned with the theme;
- the display must inform, raise questions and stimulate interest in the theme;

- display organizers must be able to explain why they chose particular pieces of work to be displayed.

In between these extremes comes the task of developing pupils' sense of what makes a good display or poster, so that they can use such knowledge when trying to develop their own explanations of their scientific work for others. This might be based around adverts in their language work, or around analysis of a display or poster which the pupils feel is a 'good' one. Perhaps a good time for this is after an initial foray into developing a display to communicate what they have done in problem-solving. AOTs — parents, local businesses, publicity firms, governors — are all useful sources of help here in assisting pupils with the skills of seeing what makes a good visual communication.

Scale drawings and associated skills are an aspect which the development of technology and mathematics in the National Curriculum will (hopefully) enhance in graded stages — and we feel advice is best left to specialists. Modelling would seem at first glance to fit into this category too, but for us, modelling has a further meaning. In science it is hugely important, being one of the major mechanisms by which pupils make representations of their environment as they understand it. However, modelling in science is also a very specialist area, one we think lies outside the realms of this book.

Scientific Writing

All pupils are used to having stories and poems read to them from their earliest times in school. It is one of the classic ways in which teachers achieve a number of things — encourage listening, develop feelings of the class as a group based in experiences of the family, extend pupils' interest in reading and books, develop a sense of drama and anticipation for pupils and develop an appreciation of different kinds of language, literature and writing. It is primarily the latter which concerns us: we are of the belief that, since science is a mandatory part of every pupils' experience in school from 5 onwards, reading to pupils should also encourage their interest, appreciation, fascination, even love of science. Of course some of that could be managed by reading science fiction stories, science biographies, or human stories of science from time to time, but it could also be done by reading poems about science, by reading, as pupils get older, scientific accounts — Raymond Dart's description of cleaning the baby Tuang skull; Mary Leaky's description of the discovery and appearance of the Laetoli footprints; Darwin's log; Watson and Crick's story; Galileo's life; Michael Faraday's inventions; Joule's honeymoon — all make interesting reading and include a variety of styles of scientific reporting. Some of the popular versions of these tales of science are more myth and magic than they are factual account, but science never lacked a little romance in the telling. This helps to focus pupils' attention on what scientific reporting is, was, and might be and assists them in paying attention to the way that they write up their own observations and accounts of work done.

A Final Word

Collaborative working in groups and teams and developing roles is difficult to teach. We have covered some of this in our earlier sections and through our progression through Years 1–6. Though it is difficult to teach, we are firm in the belief that direct skill teaching must happen if pupils are to grow in their team skills. Much early practice should happen through games and paired work, with a growing awareness of the responsibilities towards other pupils which needs to take place.

Chapter 6

Communicating Achievement: Assessment in Science

Introduction

The field of assessment has undergone a transformation since the introduction of the National Curriculum, assessment in different forms is one of its central tenets. Like all forms of assessment, the National Curriculum Standard Attainment Tasks (SATs) are open to criticism on grounds of validity and manageability (at the very least), and of being indicative of the political shortsightedness that accompanies the 'return to basics' movement. This chapter deals with none of that. It is not a guide to conducting SATs in the classroom or laboratory — others (the National Foundation for Educational Research (NFER) for instance) are better placed to advise teachers on that. Moreover, like all transformations in the happening, firm statements are difficult to make since changes are occurring rapidly, even as we write. Nor does it seem that a 'steady state' or stable position will be reached for many years to come, so we do not attempt to speculate about how SATs might interface with other assessment procedures, for example the General Certificate of Secondary Education (GCSE). In this chapter we address what we see to be the essentials: continuous, formative (sometimes summative) assessment by teachers in the classroom, and the communication of the achievements and progress of young people. That is, good assessment starts in the everyday practice of teachers, it has not arisen because of national testing procedures.

There are a variety of acts of communication within processes of assessment. At one level the chapter deals with ways of assessing the kinds of communicative skills discussed elsewhere in the book. By far the greatest part is devoted to the acts of communication contained within assessment. Teachers need to communicate with learners — to tell pupils what is expected of them, how to improve, assess their strengths and note their difficulties. Pupils need to communicate with pupils, sharing understanding and expertise and providing feedback. Teachers need to communicate with other teachers, informing them about children's progress, and — in turn — accepting information given to them. They need to gauge how much and what kind of information is really useful within school and between schools and, in turn, how and what to communicate to parents, employers and the public at large. Schools also need to communicate through public channels: for example, results are to be published and no doubt people will judge schools by these means — both by the medium and the message. Exactly what is published, how, and when, about a school's progress and achievement ought not to be left simply to chance or government decree, schools must exercise some control over the communication processes.

In this chapter we look first at assessing classroom skills, and we illustrate some approaches through a number of activities. The remainder of the chapter looks at the information commonly available from assessment and how it is reported. In this, we try to de-mystify the process and present pragmatic suggestions for dealing with actual assessment on a regular basis, managing it in the classroom, dealing with and streamlining the ever-increasing burden of administration connected with it. Third, we take a hard look at issues of communicating assessments — to pupils, parents, community and press. Last, we draw on work in earlier chapters to explore a model of assessment featuring pupil autonomy and progress in achievement.

The Story So Far

The Schools Examination and Assessment Council (SEAC), have noted assessment arrangements for Key Stage 1 (SEAC, 1990) as follows:

> ... teachers must (by 31st March each year) summarise the outcomes of assessments made during the key stage, in order to determine the level of attainment achieved by each pupil in relation to each attainment target.... By [the end of the summer term] schools must have completed, in conjunction with the LEA, a necessary resolution of discrepancies between the outcomes of the SAT and teacher assessment. At this stage, schools should have aggregated the final attainment target scores for each pupil to yield profile component and subject levels and reported these profile and subject levels to the parents of individual pupils.

The Major Points

1. During the key stage:
Teacher assessment must take place frequently as pupils are taught particular programmes of study. The level at which each pupil is achieving is recorded regularly and these assessments are used to produce a 'score' for each pupil in each attainment target.

2. By the end of March:

All attainment targets should have been assessed to obtain scores for each pupil. The scores for all the attainment targets should be summed and used to produce one score for Profile Component 2. This is a time-consuming business, especially at key Stages 1 and 2 when English, maths and science have all to be assessed by the same teacher. Some will prefer to complete assessments by the end of February, leaving a month for tidying up loose ends.

3. By early June in the summer term:

SATs are completed in English, maths and science as prescribed.

4. By the end of the summer term:

Any discrepancies such as anomalies between SAT scores and teachers' continuous assessment scores are checked, and where necessary referred to the Local Education Authority's appeals procedures.

Producing a Profile Component Score in Science

Since, for science, Profile Component 1 is also New Attainment Target 1, the level at which pupils achieve can be entered directly as the score for Profile Component 1. For Profile Component 2, the scores for each of the other new attainment targets need to be collated to produce one final score. The DES (DES, 1990) have produced a circular to assist with this calculation — ideally much of it is a computer task, and the Association for Science Education and other organizations already have such things in hand. No doubt, now the number of attainment targets to be assessed in science is reduced (Clarke, 1991), the calculation will need to be different. The final scores, for science is an average of the two profile components, though all three scores, science and profile components are reported at the point at which reporting becomes necessary.

However, the question remains of *how useful is this information once aggregated in this way*? What does it really say to parents, pupils and teachers about a pupil's ability in science? Teachers in the class to which the child is progressing need to know more detail, the un-aggregated information. Aggregated numbers may satisfy some parents, but clearly not all. Some will inevitably ask: What (for example) does a score of Level 2 in science mean? Is that 2/10? What did everyone else in the class score? Parents will need to be fully informed in a variety of ways and are entitled to have such information about their child, though not, at Key Stage 1, *entitled* to know about the level of achievement of other children of that of the school as a whole. The school may choose to publish whole school results at this key stage but it is not obliged to do so. Parents will of course be keen to learn how their child is faring in the 'basics' — reading and written work and spelling — all different attainment targets within National Curriculum English. We need to inform them that science is now also a 'basic' (a core subject) and there are important questions to be asked about pupil achievement in this too. If parents ask the right questions about achievement in science, its importance in the eyes of pupils and thus the educational world will continue to be enhanced.

Similarly, other teachers will need full but complex details when pupils move to secondary schools, at or around Key Stage 3. A simple number may be sufficient for a school 'record' but the information will also have to reflect the needs of the audience, the head of year, head of department, subject teachers, tutors. The questions remain. How can the information best be presented to take account of the varying needs? What about evidence for judgments? What about other comments?

Dealing with Discrepancies

Teacher assessment is not just in science but across other core areas as well, though it is the SATs which will ultimately determine the final pupil level. For NAT1, since it is also a profile component and to be tested by SAT, the score on the SAT will be the level at which the pupil is said to be operating. For the other profile component, SATs will not be undertaken for all ATs. Thus the teacher's judgment could carry the day.

There will be some discrepancies between the teacher's assessment and the score which some pupils achieved on the SAT. As regulations stand, these discrepancies can either be ignored, or challenged. If they are left unchallenged, then the SAT score will be the one which stands. Testing agencies always try to minimize challenges to pupil scores; SEACs expectation is that most of the discrepancies will be resolved by discussion between a visiting moderator to the school and the teacher (SEAC, 1990). However, *'resolved'* can have a variety of meanings. On one hand, it could mean that teachers accept the moderator's judgment that the SAT is the final arbiter and is thus taken as correct. Or it could mean that teachers present several assessments with accompanying weight of evidence, inviting the moderator to re-assess the evidence from the SAT and the continuous assessment. If this latter course is to be successful schools will need to establish careful internal procedures about collection of evidence. However, at the end of the day, the assessments can only be as good as the instruments used for making them — the tasks — whether they be SATs or teacher's own assessments. If a combination of these tasks cannot differentiate sufficiently between pupils' abilities and achievements, especially at the top and bottom of the possible range of levels, then moderating discrepancies will be time-consuming and difficult.

Assessment by Teachers

SEAC have provided some comprehensive and useful advice in a variety of forms. There are the teacher assessment packs (SEAC, 1989), of which pack C is a useful source of information, pack A is (in our view) unworkable and patronizing, and pack B too brief. They have made some statements about what is expected in SEAC, 1990 (*op. cit.*):

> Since assessment is an integral part of teaching and learning, the intended learning outcomes through appropriate classroom activities should be planned in advance and judgements of pupil's performance made in that context. Throughout the key stage, therefore, teachers will be making assessments of pupil's performance against the attainment targets and in relation to their defined statements of attainment as part of their normal classroom practice. (p. 3)

As with all advice it needs using judiciously, to be treated with the question, 'what can we learn from this?', rather than assuming that the illustrations used are authoritative examples of good practice.

Casting Out the Myths

We start with some assumptions. First, a truism (but none the less important for that) — assessment is an essential and integral part of the curriculum and every learning experience. In much the way SEAC pronounce, it should be planned for in the work which pupils do. However, we would see the necessity for this to be somewhat different to that stated by SEAC. They see the purpose of assessment as being something which aids the teacher in pinpointing pupil progress; while we see assessment as a tool for pupils. Continual exposure helps learners control their own progress. They can learn to be skilled at assessment, and not fear it, but see it as a formative exercise which will help themselves, and others, to develop.

Second, all teachers are very skilled at assessment. It is an everyday occurrence, probably for most of the day, as they make judgments about when to intervene in an activity, how to question, when to question, which type of question is the most appropriate, how pupils are progressing, when they need help and when they are best left to persevere. We call it 'using professional judgment'. In managing the learning environment (Chapter 2) teachers make complex assessments of pupils and groups involving many variables and make considered judgments about how best to deal with a situation or learning experience in seconds. They 'weigh it up' and make a decision, each decision an act of assessment in its own right. Assessments, then, take place all the time.

What is different about the National Curriculum and assessment? In our view, substantially nothing. It merely asks teachers to do what they commonly find most difficult — to make explicit their judgments and the process by which they are reached. In some respects the National Curriculum makes it easier for this to happen because it provides a framework for judgments, for the levels expected from the pupils. The very open question, of course, is how well the levels reflect any basis within the reality of classroom life or children's learning, if there is reflection there at all.

Preparing for Continuous Assessment

There are two levels of preparation. The individual classroom level, and the whole school level. We deal first with assessment at classroom level.

Assessment in the Classroom

One item of equipment not necessarily needed is a clip board! However, there are some prior conditions to establish before assessment can run smoothly, and we examine these first before dealing with the assessment of actual skills and looking at pupil's work. The first consideration is that of classroom organization.

Classroom organization
Good preparation is vital at all key stages. Assessment should be a normal feature of classroom life and must be planned into the programme of work, particularly for NAT1. Too often continuous course work assessment in other spheres, for example at GCSE, has deteriorated into 'mini practical tests' in many schools: days when half the class would be assessed while the other half busied themselves — invariably — with written work. Assessment was conducted by giving pupils a work sheet containing instructions which required activities under conditions very similar to those of an exam. Teachers asked questions, probed and prompted, all the while reminding pupils that asking for help might penalize their mark. In other words, a situation where stress could easily mount — a far cry from the usual lesson format or teacher-pupil relationship.

Primary teachers have the opportunity to establish a different tradition. They are familiar with organizing group work on very different tasks within any period of time, thus allowing them to decide who will be assessed, when and on which activity. To assess successfully with the minimum amount of disruption to class life and teacher work load, it helps to assess small groups. Since we believe the point of assessment is the achievement and progress of the pupil, then the degree of pupil autonomy the classroom offers is paramount.

One way to decide if classroom organization is fully geared to assessment is to conduct an 'organization audit' and the checklist below is designed to help. Some of the questions apply specifically to secondary or primary schools, though we believe that the majority apply equally to all classrooms.

AUDITING CLASSROOM ORGANIZATION
You intend to assess small groups of pupils working at any one time, in conditions that give you time to spend with them. So,

Do the physical arrangements enable group work?
For instance:
— does the arrangement of tables/desks facilitate or hinder pupils holding discussions in groups?
— are laboratory work stations clearly labelled?
— can pupils move around easily without disturbing others?
— are basic materials for classroom/laboratory life clearly labelled and accessible to pupils?
— are there areas of the room where different activities can take place, such as reading, working on a piece of quiet work, getting on with a project etc?
— can you see all the areas of the room wherever you are standing?

Do pupils operate autonomously?
— how often are you asked for materials which are easily accessible in the room? what is your response?
— how often do pupils sit with their hand up waiting for your attention?
— do you have a queue of pupils waiting at your desk? What is the average length of time spent in the queue?
— how often do pupils use their peers as 'experts' who can help them solve a problem? Do you encourage this to happen by your response to questions?
— are sources of information (dictionaries, reference books, reading books) easily available in the classroom?
— do pupils have responsibilities to the whole group for — say — keeping trays, cupboards tidy?
— is there one designated to take responsibility for the whole room, such as a class 'chairperson' who ensures tidiness of chairs or acts as a spokesperson?
— what is your expectation for pupils who finish a piece of work? To read? To talk to friends, complete other unfinished tasks? How do you prepare them for this expectation?
— how do pupils decide when to move on to a new activity? Do they ask you, does the group decide?
— do the pupils remain on task only when your attention is on their group in particular?
— how much of your time on average does each pupil receive in a time span (lesson or morning session?)
— how often do pupils go out of the room to use other parts of the

school environment (the library, the grounds, the computer) as a normal part of what they are doing? How do they seek permission for this?
— what expectations do pupils have of helping peers who are having difficulty with a particular idea?
— how often are pupils asked to account to the teacher for time spent during a period, to explain why they chose to complete a piece of work or practise a particular skill?

Does the curriculum preparation draw on pupil's autonomy?
— do pupils know what they have to achieve in the time span? Are the day's or lesson's tasks written for everyone to see?
— how often are different types of work planned to take place in a session (discussion, writing of previous work, practical work, surveys, library work, preparing a report)?
— what role does pupil self-assessment and peer assessment play in your teaching?
— what role does peer teaching have in the classroom?
— do the pupils know exactly what the goals of the lesson or piece of work are?
— how do pupils record their own progress through tasks over time?
— how often do you prepare practice exercises for pupils who feel that they lack confidence in aspects of what is being done? Can they practise areas of self-assessed weakness easily? Frequently? where do they go to check answers or procedures, to the teacher or peers?
— how often is marking used to direct pupils towards practising particular aspects of work, and what check is made to ensure they do this?

Planning
One way in which the essential differentiation required by the National Curriculum can take place is to plan the curriculum carefully. In order to design an effective 'planning unit' then, classroom organization needs to be considered. In Key Stages 3 and 4, for example, a 'planning unit' usually represents a six week, half-term period, and time is deliberately built in for practising skills such as hypothesizing, interpreting data, using particular kinds of measuring instruments. Periodically then, pupils, through a teacher's marking, can be directed to practice skills. They might read, develop different communication skills — such as redrafting work in a different register, writing an article on an experiment for a popular magazine or a scientific journal, or taking data and re-interpreting it in terms of a pie graph using a computer. From time to time, this would mean that groups in the room would be engaged in tasks which would allow time to be spent with them, assessing and practising skills without the need to orchestrate the whole class all at once. It

would also give pupils much more control over their own learning, allowing them to develop confidence in particular areas.

Many teachers at Key Stages 1 and 2 already operate in this way, and as such will feel the burden of assessment in the classroom much less their colleagues.

We noted earlier that planning for science in the National Curriculum needs to take place over a whole key stage, rather than each attainment target being delivered each year and we return to this issue in a later section on whole school planning. Planning activities with assessment in mind suggests a flow chart as follows.

It is of course possible to assess in ways that are very restricted in range and allow assessment of only one level in one attainment target. It may be necessary to do this if a child has been absent, or there is a particularly difficult concept to assess which becomes clouded by other activities. Given the amount of time available for assessment and teaching, we feel it is much better to plan activities to enable the teaching of the planned curriculum and then adjust these to provide a greater opportunity for assessment.

In the flowchart, we have suggested that activities in science can be planned and then adjusted to enable assessment in other subject areas. Clearly this is an advantage for teachers in Key Stages 1 and 2 who have a greater range of attainment targets to assess. We want to repeat, too, that we assume assessment is an everyday activity for teachers, and that each teaching task provides the potential for assessing and gaining clues about a child's level of attainment. Key Stages 3 and 4 teachers need to be less concerned with the assessment of other subjects, but do need to be concerned with the range of levels in their own subjects which can be assessed through an activity. The example below is for Key Stages 1 and 2:

Example of Assessment for Key Stages 1 and 2

A Key Stage 1 teacher gave a small group pupils a set of sponges (natural and non-natural), corals and shells and asked them first to group the objects in categories or sets of their own choosing (they chose hard and soft then bumpy and smooth) then in categories of her choosing (she chose natural and non-natural). During the activity she explored with them the idea that some items fitted into both sets — or rather were on the edge of one, some shells for example being smooth on one side and bumpy on the other. She was able to assess the pupils' achievement of NAT4 levels 1 and 2a with this activity. She also reminded them about 'shared' sets and the pupils re-adjusted the plastic set barriers she had provided to make three inter-locking sets. From this activity, she could assess pupils' understanding about natural and non-natural materials (NAT4 level 3) although not in full, since the non-

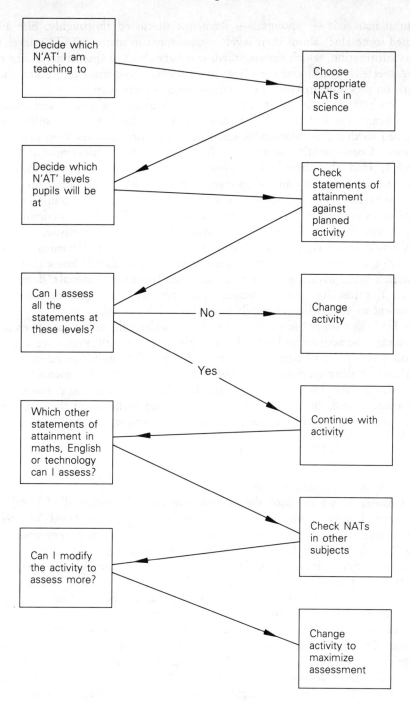

Figure 6.1: Flow chart showing planning activities and assessment potential

natural materials — sponges — were not discussed thoroughly. She also gained some clues about their level of attainment in maths (OAT10, level 1). This information, which she recorded, was later checked against her maths record sheets. It confirmed her earlier assessments about most pupils and their work on sets and provided extra information for one pupil.

She extended the activity by asking pupils to look at a book and classify the sponges she had provided. Unfortunately the book was not sufficiently detailed so that classification required nothing more complex than matching pictures. Consequently she was not able to assess pupil's achievement of NAT4 level 3. Had she altered the activity a little and asked pupils to make comparisons between pairs of sponges — perhaps by a flow chart, she could have assessed this level since they would have needed to make comparisons of similarities and differences much more thoroughly. The activity as defined for pupils was not helpful because the classification was too simplistic. Even so, this would only have provided a clue as to the children's achievement at level 3, since the actual statement of attainment says 'pupils should know that some materials occur naturally while some are made from raw materials. Being able to do this just for sponges would have provided some evidence, but not sufficient to state whether pupils had achieved this level.

Had the teacher been provided with a flowchart she might have been able to change the activity and extend it to technology as well. Pupils were asked to observe a piece of coral or sponge in detail and draw it. Several measured the sizes of their specimens and looked at it under binocular microscopes or hand lenses. At this point the notion of 'what made the holes' could have been introduced, and pupils asked to design and make, from their observations and measurements, a model of the kind of animal that had lived in and built the coral.

Assessing communication skills

In Chapter 3, we outlined the communication skills required of pupils in science and in this section we explore how these might be assessed. We stress we are not making statements about the assessment of skills as they appear in the English National Curriculum. Assessing science activities alone cannot provide sufficient information for assessments in other secondary areas to be made as well. However, Key Stages 1 and 2 teachers can use activities which allow them to assess science whilst picking up clues about the child's attainment in English and maths as we illustrated above.

For science, we start with the first 3 levels of NAT1 and draw out those points first which can be assessed by written evidence, and then those which require the teacher's presence with a group or an individual. From level 4 onwards a great deal of evidence will be collected from written communication alone. However, considerable involvement with groups of pupils will still be needed to observe and assess other science skills in action. As far as communication skills are concerned, these are opportunities not to be missed.

Communicating Achievement: Assessment in Science

Attainment Target 1

Level	Act of Communication	Possible source of evidence or assessment activity
1	— describe and communicate observations	Communication here is verbal and teachers will need to listen to pupils talking in groups or telling another adult what has been observed.
2	— ask questions — suggest ideas	Here evidence can be written, particularly if pupils give a fairly graphic account of what they did. Often though, the teacher will need to note when pupils in a group begin to ask questions or set up their own investigations without asking questions. Activity 1 below gives an example.
3	— formulate hypotheses (make four tests)	Increasingly, as children develop, evidence can be obtained from written sources but frequently hypothesizing is missed out because pupils do not always commit these to writing. Since there are seven statements in this level, it will take quite a time to accumulate the evidence for all. It will help to observe pupils in action in a group, or to teach them how to recognize a hypothesis when someone else makes one.
	— interpret observations	This can be assessed fairly well by written accounts of work.

ACTIVITY 6.1
ASSESSING LEVELS IN AT1 DURING AN ACTIVITY INVOLVING 'FINDING OUT ABOUT SHINY THINGS'

This is an activity which was carried out with key stage 1 pupils. We detail the role played by the teacher and the conversations she initiated or overheard, which helped her make judgments about pupils' levels of operation. Much of the dialogue by pupils Paul and Jenny has been omitted. The italicized script represents our asides on the evidence as it presents itself, though our judgments have not been moderated and their progress may be less definite than we portray here.

In the activity pupils are provided with a selection of shiny materials: mirrors, coloured filters, torches, shapes cut out of black sugar paper and similar shapes cut out of coloured filters — shapes of rabbits, stars, circles etc. They are asked to find out which materials 'let light go through them and which don't'. They work in pairs or alone, and continue for as long as they like. They are asked to draw what they see and to tell the teacher — or the welfare assistant — when they see something really interesting. Teacher makes sure that across the half term she listens and talks to each pupil about the activity, which is left set up on a table at the side of the room.

Jenny and Paul work side by side, each with a torch and the shapes. As Jenny shines her torch at the shiny paper, she notices that she can see

the round circle of the torch in the paper. The black cut out rabbit doesn't do this, and she tells this to the teacher:
Teacher: are there other things that we can see the torch in Jenny? Jenny successfully finds two or three other shiny materials, and one red piece which looks shiny but actually lets the light go through to the other side. She holds it up to her face and looks through it. She says the world is now red. She tells teacher which shiny things let her see the torch.

Without knowing the word 'reflection', Jenny is beginning to grasp the idea that some substances 'stop' the light (the shiny paper and the black rabbit) some let it go through, (the red shiny paper) and some reflect the light back to her eyes (the shiny paper) She is beginning to have some understanding of level 2 in NAT5.

The teacher encourages her to look for other materials that she can see through and which make everything a different colour. She chooses the coloured filters and shines her torch through them.
Teacher: (holding up a green filter) What colour will the light look if we put this in front of the torch?
Jenny: (who has used this filter it already) Green.
Teacher: What about this one? (Holds up a magenta filter which Jenny has not used)
Jenny: Purple

Jenny is predicting here. Her vocabulary does not stretch to 'magenta' so to her it is purple. Nevertheless she is still predicting. She is suggesting why something happens, so may be beginning to work at level 2a in NAT1. However, more evidence would be needed to be sure that she has fully grasped ideas of why the light passes through some objects, reflects off others and is stopped by still others.

Paul, meanwhile, has been exploring the solid shapes and discovered that he can make shadows of the different shapes on the wall. He tells teacher about the rabbit and triangle shadows.
Teacher: Where do you think the shadows came from Paul?
Paul: I think the shape made it.
Teacher: How do you know?
Paul: Because it's the same as the shape.
Teacher: Hold your rabbit shape up to the wall and turn off the torch. Can you see a shadow?
Paul: No
Teacher: So did the shape make the shadow?
Paul: No there's no shadow if there's no torch.
Teacher: How do you think the shadow got there?
Paul: The torch made it.
In this sequence of questioning Paul is moving towards being able to form a hypothesis — 'I think the shape made the shadow because...'.

> *He is certainly trying to explain why the shadow has the shape it does. For Paul, a possible level in NAT1 is 2a. However, although he can associate the shape with the shadow, he has not understood that what the shape does is to stop the light to create a shadow.*
>
> Paul and Jenny begin to talk together. Paul shows Jenny his shadows. She likes the rabbit, but wants to make it purple, so she tells Paul to put the magenta filter in front of the torch.
> **Paul**: Why?
> **Jenny**: Then the rabbit will be purple. It makes everything go purple — look, hold it up and look through it.
> Paul puts the torch, magenta filter and rabbit shape in the right order.
> **Paul**: Hold the rabbit.
> **Paul**: It's not purple — it's still black.
> **Jenny**: Why isn't it purple? When I look through it it's purple!
> *Paul is clearly operating at level 2a here. He asks Jenny why the rabbit will be purple. Jenny too, by her response shows that she is operating at this level. Her statement is one of the 'what will happen if' variety, but phrased as a statement. It clearly shows moving towards a hypothesis, even though based upon false premises and her present experience of light as it appears through a coloured filter.*

The teacher's role here is clear, she is acting as a questioner and an enabler. However, without her involvement in the activity much of the evidence for development would be lost. Much of what the teacher will record at this point will be tentative. She has indications that Paul and Jenny have operated at level 2a in NAT1. She will need to listen and work with them when new activities take place based on different targets to be sure that they really are asking 'why' and 'what would happen if' — or doing the activities associated with 'what would happen if'. If Jenny and Paul had not been together, they may well have carried out the 'making the rabbit purple' activity without saying anything — another excellent reason for group or paired working: it gives children the opportunity to make explicit the things they are thinking. If Jenny were working alone, the teacher would have to use her own questioning skills to enable Jenny to share her ideas about the rabbit.

Assessment at Whole School Level

Before serious assessment by individual teachers begins, some aspects at school level need careful consideration. These involve communication with colleagues and pupils about assessment. Most of our examples below apply whatever the age phase, but where particular issues arise we have tried to signal and deal separately with these.

Creating a whole school assessment policy
SEAC (SEAC, 1989) have some advice for what such a policy might be — it contains information on:

- how progression is to be organized in the school through schemes of work (planning)
- how and by whom continuous assessment is to be monitored (management)
- arrangements for making sure teachers are interpreting levels of achievement consistently — both in their own classrooms and across more than one class (moderation)
- recording achievement in the National Curriculum and other aspects of school life (recording)
- recording and storing evidence of pupil progress (evidence)
- how and what information is to be passed on from teacher to teacher about pupils' progress (reporting)
- how and what information is to be reported to parents, the LEA and the outside community about pupils' progress (reporting)
- how statemented pupils are to be dealt with (Special Educational Needs)
- how disapplication of pupils is to be dealt with
- sharing the process with the child.

(adapted from: *A Guide to Teacher Assessment, Pack B*, p. 3, SEAC, 1989)

We feel that there are two essential aspects missing from this policy statement:

- pupil responsibility for progress and assessment
- assessment to support pupil progress.

One might see these two aspects as being included in SEAC's final note on sharing the assessment with the child, but nothing in their material leads us to suppose a formative and collaborative process between pupil and teacher, in which each might have a valid though different view of progress. It is portrayed very much as a 'teacher directed' model.

What might a 'communicating achievement' process look like? Below, we have tried to depict this for a primary context, but it applies equally as well to secondary schools.

THE WATTLEY ACADEMY FOR AUTONOMOUS LEARNERS
COMMUNICATING WHAT LEARNERS ARE ACHIEVING
A POLICY STATEMENT

Planning
Teachers and pupils have a joint responsibility for recording progress

towards set goals. Pupil statements of the National Curriculum attainment levels have been prepared by the individual subject coordinators and are available for all pupils in all subjects. These are available from the school secretary on request. Similarly, self-assessment records of progress and practice are available in the same way.

The school has agreed that planning sheets and schemes of work will show how progression, differentiation and mastery is developed for pupils during each half term.

The aim
Assessments will not take place during the first half term with a new class so that teachers and pupils can get to know each other well. Assessments will then take place regularly and frequently. It is expected that every pupil will be assessed by teachers on some area of the curriculum each week, that all pupils will assess themselves twice a week, and be assessed by their peers in different groups once a week. Half termly discussions will take place between pupils and teachers on the work each has selected as indicative of a particular level of attainment. Pupils and teachers together will decide when a piece of work has been superseded by a higher level of attainment and pupils can then take examples of work home for parents.

Moderation
Once a term, the deputy head will work with each teacher in joint assessment tasks in the classroom, to enable some internal moderation on areas that require other than written or physical outcomes. Internal moderation meetings will also take place each half term between teaching teams and the Governors' working group on assessment. The deputy head will also collect examples of tasks set, and the assessments made, for the 'Assessment Bank'. Everyone is free to use the bank regularly for help in assessing. We have made arrangements to share the bank with St Sams, who are establishing their own, similar, process. We are in the same moderating group as both St Sams and the Rhiannon School, who are invited to team moderation meetings.

Collecting evidence
Storage cabinets have been installed near the staffroom for the collection of work, to leave as much space as possible for children's activities in their classrooms. Pupils will know about collecting evidence of their achievements, and the part they play in this. Parents can send in examples of work the child has done at home and in other places, and we encourage parents to keep their own record of achievement at home using our recording forms and coloured 'level' stickers. The school camera is available should staff wish to record really unusual work.

However, tape recordings and photographs should only be used to capture otherwise ephemeral evidence. Otherwise, seek the help of a colleague to sort out interpretations of a particular level.

Reporting
The 'Explanation' sheets on the National Curriculum have been very useful in reporting levels of attainment to parents, and parents are now free to ask for examples of how pupils are progressing in different aspects of the curriculum. The report sheets are available from the school secretary for staff and pupils to complete. These will be sent home once a year, while regular comments on achievement in pupil's diary books will keep parents well informed between times.

Marking
All marking should reflect progress in the National Curriculum. Pupils' half-termly planning sheets detail what we intend to do, and which attainment targets are being worked at. Marking should draw attention to success in the goals set. For example where written work shows evidence of hypothesizing in science or making an interpretation, this should be commented upon to the pupil. Teachers of older pupils may wish to prepare them for Key Stage 3 experience by giving marks. If so, marks should indicate the different parts which reflect our agreed principles for each subject.

Whole School Planning

Although some aspects of planning are included in our (fictional) communiqué on assessment above and we have dealt with it in detail in the previous section, there are some features which need more exploration. Planning pupil experiences is a collaborative venture between teams of teachers in most primary schools and within departments in most secondary schools. However, the planning may need to be different for each key stage. In Key Stages 3 and 4, for example, it would be very unlikely and unreasonable for all seventeen attainment targets to be taught every year. The resulting time devoted to each would be so small that pupils would gain little from the experience. It then becomes pertinent to ask: if an attainment target is taught in year 8 (12 year olds) and not met again in that key stage, will the assessment of that pupil's understanding still be valid twelve months later? We must make such an assumption at Key Stage 4 when pupils take GCSE, because it is a two-year course and teaching is serial throughout — topics cannot be fully repeated in the second year. The teaching once only in a key stage then raises the spectre of revision sessions before assessment of attainment targets on topics taught many months earlier.

Communicating Achievement: Assessment in Science

In a nutshell, teaching topics across a two-year span (as we do presently at GCSE) or across a three-year span (as we may choose to do at Key Stage 3) implies that pupil progress and understanding slows down with age, or somehow becomes frozen at one level, and only changes when we teach or re-address it. As all teachers know, the realities of learning are different. Understanding will sometimes regress and pupils cannot later use key concepts and skills they once grasped. Though they might understand that water, light energy and carbon dioxide are needed for photosynthesis (NAT3, level 6) at the age of 13, there is no guarantee that they will still recall this when they are 14. Some ideas advance and are enhanced through other areas of work and experience. So, for instance, their concept of energy will change and develop as they encounter this in other parts of science. Thus, if they are assessed as achieving level 6 on the basis of continuous assessment when NAT3 or NAT5 was taught, is this acceptable information to hand on to teachers at Key Stage 4? We believe that, if pupils are given some power over their own progress, then such problems can be diminished. We tackle the practicalities of this later in the chapter.

Planning, then, for teaching and assessment opportunities has several stages:

1. Whole school planning — field trips, whole school activities, school themes, school tests etc.
2. Year team/departmental planning — joint planning for the classroom activities to achieve the teaching of the attainment targets in the required time available.
3. Individual planning by the teacher — planning of activities which pupils will do in the classroom and which are designed to provide the pupil with the maximum opportunity to learn and use their learning and to provide the teacher with the maximum opportunity to teach whilst at the same time finding out where pupils are in their own achievement.

In other words, for us, the bulk of the time available in the classroom should be for teaching and learning. Assessment should be planned into the teaching as an opportunity, not distort it or replace it.

Conducting Continuous Assessment

Evidence

There is one major question that needs to be addressed before we consider the notion of evidence. That is who is the evidence for? If the evidence is for us, the teacher doing the assessing, then any range and type of evidence is open to us. It is simply the basis for our professional judgment. We can use all and any observations, conversations, experimental situations and so on. If the evidence is for others, we need to consider the recording of the evidence in order to preserve it. This is obviously not a problem where written work or drawings, diagrams or other two-dimensional aspects are the basis for judgments. Where more ephemeral evidence is used, then recording in some way will be necessary. At this point tapes, photographs and other means may have to be considered. In part A, we deal with the evidence which the teacher will use for their own purposes.

For the assessors
Teaching produces a wealth of evidence which can be used already as the basis for judging achievement. The activity below may help to define what is already available in the classroom.

ACTIVITY 6.2
Complete the columns below to show what evidence is already available in the classroom to use for assessing progress. The first two have been completed to provide clues.

		Evidence.		
verbal	written	2d	3d	physical
discussion	poems	diagram	model	mime
debate	account of an experiment	drawing	construction of apparatus	skills

---

Clearly, this is a general activity. It will remind us of what we can use. It will not target what is the best type of evidence for any one attainment target or level. Throughout the chapter so far it has become clear that certain kinds of evidence lend themselves best to the assessment of certain kinds of attainment levels, and that often, this changes with age. Certain scientific skills, for example can only be assessed by means of observation on behalf of the teacher as the experimentation takes place.

To illustrate this point, we include here three activities which might be appropriate for science in Key Stages 1–3. We have not included Key Stage 4, since this will be assessed by the criteria of whichever GCSE examinations the school chooses. The activities are accompanied by an italicized commentary on the most appropriate sources of evidence when assessing.

ACTIVITY 6.3

KEY STAGE 1
THE TASK

The theme which has been planned is Transport. It encompasses New Attainment targets NAT1, 4 and 5 in science, ATs in maths and all four ATs in technology.

As a major part of the theme, the pupils have to make a vehicle for four people, which will not slip down hills in icy weather. In this activity, pupils have been asked to find out how they can stop vehicles sliding down hills when the roads are slippery, to find out something about roads, cars and slipperiness before they make their vehicle. The teacher has provided some smooth wooden slopes, some hardboard — the 'bumpy' side up — some oiled wooden slopes and some slopes covered with roofing felt so they are gritty. There are a variety of toy cars for rolling down the slopes, with different wheels and of different masses.

CARRYING OUT THE TASK

The children work in groups, as is the normal routine in the class, so that at any one time the teacher is only assessing her targeted group of pupils. She has asked them to try the cars on the different slopes and to find out which ones are the best on the slippery slopes first. She participates in their discussion and from this can identify that all the pupils can group materials by means of similarities and differences. They decide the smooth wooden and the oiled slopes are slippery and the other two are not. The teacher has provided word cards with the words slippery, oiled, smooth, wooden etc. on.

From the discussion, the teacher can identify which children are achieving levels 1a in NAT1. Access to this level might have been possible through 2D evidence. She could have asked the children to decide which slopes were slippery and which were not, and to draw them and write the names of the slopes alongside.

The group discuss what 'best on the slippy slope means'. The teacher reminds them that they have to make a vehicle which will not slip easily when the roads are icy or wet. The children decide that the cars that do not slip on slippery roads will be the best ones. They try out the different cars. One child (Leon), about half way through begins to insist that they put all the cars at the top of each slope and see how far they go.

The teacher has an indication here that Leon has some understanding of a fair test, (NAT1 level 3a) although he does not actually use the words 'fair' or 'test'. She can record this for verification later in another situation. It is unlikely that any of this could have been found from written evidence with children of this age.

Samantha insists that it is how fast the cars go not how far they go that counts. The others agree, and the children decide that the fastest ones will be the best ones. The teacher intervenes at this point to remind them about slipperiness and what they want to achieve with their own vehicles. They finally agree, with help, that it is the slowest car which is the best.

Clearly the whole group have a confusion about the appropriate variable required here. No-one in this group appears to be able to operate, in this situation, at level 4a or 4b in NAT1. Again, the teacher's presence is crucial at this point as a teacher, to keep the pupils on the track. This would not have been possible with written evidence.

Work progresses until the children are agreed about a group of cars which go slowly down the slope and others which go fast. No one suggests timing the cars, the 'fast' and 'slow' is a subjective judgment by the pupils. The teacher asks why they think the slow cars go slower. Some pupils say that it is because their wheels are large and or very bumpy, others have no answer. The teacher asks them to draw the groups of slow and fast cars and then say which one they would best like to ride in on an icy day.

The written evidence could have provided some of the information for assessment. It would have told the teacher that observations were possible on the basis of an experiment. It would be a record of results (NAT1 level 2b). It would show that pupils could count up to a particular number and that, for some children, variables — such as wheel smoothness etc. are being identified as causal in a situation (AT1 level 2e). However, other aspects would have been missed — the issue of a fair test by Leon, the reasons for selecting variables for slowness, explanations of slowness etc. These could only have been accessed by the teacher through observation.

The design of the activity does not lend itself to assessment by means of 3D evidence — that will come later when the models are constructed, or physical evidence, since no measurement takes place. Verbal and 2D evidence lend themselves best to this activity and this age range.

ACTIVITY 6.4

KEY STAGE 2
THE TASK
The theme the pupils are working to is 'Our Town'. Part of the work involves looking at the local river and finding out about the animals and plants that live in and around it and if it is polluted. Pupils have visited the river. During the visit, they drew a map of different parts, above and below the road bridge. They took the temperature of the water, with the help of adults, and they photographed swans and other birds living near and on the river. They collected water samples, plant samples from inside and around the banks and tested the flow rate with sticks which they dropped from the bridge. During this activity, they worked in fours with an adult. The teacher provided a checklist for the accompanying adult to check when children made different observations. From this she has a great deal of information which may help her in assessment, but she will have to ask questions of pupils when they are working in the classroom to verify some of what she has on her checklist.

CARRYING OUT THE CLASSROOM ACTIVITY
When the pupils return, they are asked to look at the specimens brought back and examine them with hand lenses in attempts to identify them. They also have to put some of the water on a glass on the radiator to see what is left when it evaporates. The teacher walks around as the groups are working on their identification tasks and observes their discussions. The discussions help her to identify those pupils who are operating at level 3b in NAT2, and those who need more practice at being able to use keys to identify animals and plants. Listening to the pupils talking about their water evaporating and comparing water from different parts of the river allows her to know if they can achieve level 2b in NAT1. Looking at the checklist allows her to decide by talking to pupils, if they can read a thermometer to a degree of accuracy (NAT1 level 3b), understand some of the causes of pollution recorded in the backwater near the bridge (cans, oil from boats, foam). Because of the nature of the activity, little written evidence is available from the pupil's work which will help the teacher to make decisions.

KEY STAGE 3
THE TASK
The unit of work in science focuses on the environment. It involves work drawn from NATs1 and 3. This activity is designed to investi-

gate the soundproof qualities of different materials. The pupils are given a set of materials which could be used for soundproofing. They are told that complaints have been made by neighbours of a family where the child has just received a new stereo radio with speakers for its birthday. A plan of the child's bedroom and the neighbour's flat is provided (see Figure 6.2) and they are asked to put forward suggestions for soundproofing to solve the problem. A piece of plasterboard is also provided as is a decibel meter for sound readings. The pupils work in a group and are asked to keep a diary of their work; who suggested or decided what, who did what in the group. They are asked to take whatever measurements they feel are needed and write up the final findings as a report to the environmental health officer with recommendations as to the feasibility of the soundproofing on the grounds of cost. During the two lessons required from set up to completion, the teacher walks about and observes what is happening in the groups.

CARRYING OUT THE TASK

In one group, the discussion focuses around the level of the noise — what will be very noisy? The group decide upon the level which they feel would be noisy if heard through the plasterboard (wall) by the neighbours, and the critical level to which the sound must be reduced to be unheard next door. This involves discussion of 'being heard next door' — with ears pressed against the wall? with the neighbour's television on? whilst sitting reading in the neighbour's sitting room? The latter is decided upon and the group designate someone to be the tester of the lowest sound level which this might be. This involves further discussion of how far from the wall adjacent to the child's bedroom the person should sit to read, and these aspects are agreed upon. The measurements then take place using a decibel meter and tape measure.

Hearing the discussion allows the teacher access to who is operating at which level in NAT1 during the discussion. The write up (see Figure 6.3) will be agreed by the group and so the information will be less reliable about each individual. From the discussion, the teacher is able to assess different group members on their achievement of NAT1 levels 4a and b.

The group agree on how to soundproof the room, what size of materials, where to place them, and actually try out several different suggestions — soundproofing the whole inside wall of the child's bedroom, soundproofing only part of it immediately behind the stereo, using different thicknesses of materials etc.

Watching the measurements and experimentation allows the teacher to assess individuals on NAT1 levels 4a,b and 5a and b.

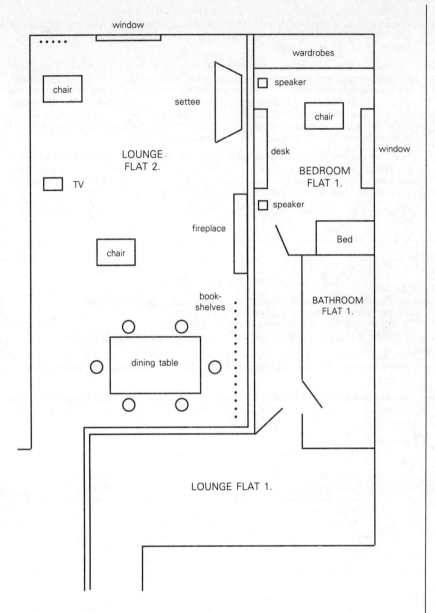

Figure 6.2: Plan of flats for work on noise

Investigating Noise and Soundproofing

We worked in a group of four. Each of us took a different job. We had to find out how to soundproof the bedroom in flat 1 so that the neighbours in flat two couldn't hear the new stereo.

There were lots of steps to our experiment. First we checked the decibel meter with different sounds and agreed how loud we would have the stereo, in all our tests. Then we agreed a scale model of the flats should be made. Two of the group made the model, using shoe boxes and a piece of plasterboard wall in between, so it would be real. We stuck the cardboard boxes to the plasterboard with sellotape, but this wasn't really very good. We couldn't solve the problem of how to keep moving the plasterboard to soundproof without keeping making a new model.

Then we put our taperecorder in the box where the speaker would be and turned it on. We put the decibel meter in the lounge (box) of flat 2 where the settee would be, because we agreed that this would be where the noise from next door would be loudest and most annoying if you were reading and/or watching TV.

When we had made our model, we brainstormed what we could do to soundproof the room.

These were our best ideas.
1. Put thick cushioned wallpaper on one or both sides of the wall.
2. Hang curtains on the wall where the desk is.
3. Cover the wall in polystyrene tiles or roll.
4. Build a box for the speakers that was soundproof except at the front.

We also thought you could move the bedroom round so that the speakers were where the bed was, but the teacher told us that the desk was fixed and couldn't be moved.

We decided to test out ideas 1, 2, 3 and 4. We split up the jobs so different pairs did different ones. We got some more plasterboard so that we could get more than 1 idea ready at a time.

Our results

Type of soundproofing	Taperecorder volume	Sound level in Flat 2
thick wallpaper in bedroom	set at 6 (80 db)	72 db
thick wallpaper on both walls	6 (80 db)	65 db
curtains	6 (80 db)	50 db
polystyrene tiles	6 (80 db)	53 db
polystyrene roll	6 (80 db)	51 db
soundproof speaker	6 (80 db)	74 db

Communicating Achievement: Assessment in Science

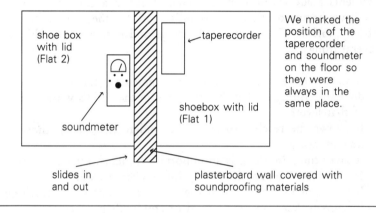

What we found out and what action we would take

We found that curtains were the best. We used velvet for ours and hung a thin lining inside. It wouldn't be so good if the material was thinner. The next best was polystyrene in a roll. Soundproofing the speaker and putting wallpaper in the bedroom wasn't much good. The owners of flat 2 would have to use thick wallpaper too, and they might not want to because it is expensive.

We would suggest that the people who live in flat 1 should put polystyrene roll on the wall. Curtains would be best, but they would have to be thick and this would be expensive. Velvet costs £8.00 per square metre. Polystyrene only costs £10.00 for a roll that would cover 10 square metres. Even with the glue needed this is cheapest.

Diagram of our apparatus.

Figure 6.3: Workcard on soundproofing and noise

When the group have finished collating their results, they choose to do individual write ups, interpreting their results as individuals, since some disagreement has developed about recommendations on the grounds of cost.

The teacher can use the written evidence to assess NAT1 levels 4a,b and level 5a. Further work will be needed to assess some of the parts of other attainment targets. This could be a piece of homework or class written work based around the results of the investigations, exploring some of the more theoretical aspects of sound such as frequencies, and amplitudes of vibration. Or, if the teacher preferred, a more constrained investigation could be set up, asking the pupils to vary the frequency by specific amounts and the amplitude of vibration to investigate the effects on their 'best' soundproofing materials.

Throughout the activities, it is predominantly verbal and 2D categories of evidence which have been used for assessment. In experiments where pupils are asked to set up an experiment according to instructions, the actual apparatus itself can be used (3D evidence) to assess aspects such as. Physical evidence is used to show the skills of measurement (as in the KS2 and 3 activities) such as NAT1 level 5c.

Evidence for others

In the activities above, the balance of the evidence tipped in the favour of written as the pupils became older. Having written evidence available is a means of security for teachers, because it provides the opportunity to preserve the judgments made about a child's achievement at a particular time. Some teachers have begun to write on pupil's work as they mark it, the level indicated by various statements or drawings and dating them. This method does several things.

1   It allows the teacher to check her judgments again later. For example, does she still agree that the statement in the child's work is indicative of that level?
2   It allows the teacher to check her judgments against those of colleagues easily.
3   It saves time. Marking is an important feedback activity for pupils. It can now be a feedback activity for teachers as well.
4   It makes assessment part of the everyday processes of teaching.
5   It helps to keep an easy record of children's progress.
6   It can be used to show where help was needed to achieve the level — recording with a comment the teacher's question which helped the child to understand.

Some of the other types of evidence, such as the verbal or physical cannot be so easily preserved. Tape recording can achieve this for discussions, but we need to sound a word of caution here. If the purpose of the recording is as a planned part of the work, which the teacher will use, or if it helps the teacher to assess because the groups are large and she cannot be certain of reaching them all, then this may well be a legitimate use of the tape recorder. If however, the evidence is being preserved for the sake of preservation, then we would submit that this is not necessary. There will be sufficient written evidence to provide a balance for moderation should the teacher's judgments need to be checked. It is not necessary to preserve tapes just to allow for checking of judgments.

Similarly for physical evidence. Videotaping is not really feasible in classrooms. It is costly, managing the collection of the evidence is time consuming, and unless it really is the only way to allow the teacher to make judgments — for example in drama — should not be undertaken. However, it

Communicating Achievement: Assessment in Science

DATE _____

ACTIVITY: Speeds of dissolving of alka seltzer depending on how broken up the tablets were and the temperature of the water.

ASPECTS LOOKED FOR	LEVEL
Variables identified:	
temperature) independent	3
size of lumps)	3
time to dissolve dependent	3
Plans to alter variables correctly:	
temperature constant, lump size differs	5
lump size constant, temperature differs	5
Controls: temperature accurately	3
stirring of water	7
volume of water	7
Uses two or more tablet sizes	4
Uses two or more different temperatures	4
Uses appropriate volume of water	5
Selects appropriate apparatus	4
Works safely throughout	7
Makes appropriate and accurate measurements	4
Written evidence	
Draws conclusions about tablet size and temperature effect	4
Includes critical evaluation of experiment procedure	6
Interprets findings in terms of particle motion and surface area	7
Records results appropriately in a table	4
Uses scientific terms correctly	4
Plots line graph of temperature and time accurately	6
Produces clear report with diagram, symbols and data.	7

Figure 6.4: Checklist for new Attainment Target 1 assessments

may be useful to preserve a checklist, dated, for the experiment, to show what judgments the teacher made about the particular experiment at the time and how these matched against the levels. Figure 6.4 shows one such checklist. 3D evidence — models etc. — create a storage problem for teachers. If evidence of the models is required to be preserved, and the pupils may want it as a part of their profile, then photographs are the best solution. Otherwise, schools will run out of storage space.

Communicating in School Science

2. Interpreting Evidence

For obvious reasons, we chose a piece of written work for this section. Once work has been completed by pupils, it can be assessed. However, making the interpretations about the evidence is by no means easy, a variety of skills are involved. These include:

— being objective and trying to keep separate facts and inferences
— setting up clear criteria for what is being assessed before the assessment takes place
— being confident about what is seen and heard and how this fits into your criteria
— identifying significant points about the evidence
— recognizing the importance of variables — for example, in how the teacher instructs different groups.

The assessment folder for Key Stage 1 has several examples of work to help in interpreting evidence. Here we present only one, with our assessment and commentary below.

The context here is that the pupils worked in groups to set up an electrical circuit. Prior to this, the teacher had discussed words such as circuit so they had some understanding of the word. They wrote up the work as individuals, under the headings 'title', 'what we needed', 'what we did' and 'what we found out'. Previously, too, the teacher had discussed issues of electricity and safety and the children had made posters highlighting these points, which were still pinned to the wall. The children were almost 7 years old at the time the work was done.

New Attainment Target 1
There is some evidence that can be gained for this from the written work.

Level 1a. The activity does not ask for statements about observation except in a very limited form — a bulb lights or does not light. More evidence would be needed to say that the child could achieve this level. Level 1 is achieved — the observations actually made as required by the activity — the light still lit up, and it did not light up' are communicated.

Level 2. Levels a-c are not required by the activity. In this sense the activity is not fully exploited. Some information regarding level 2a could be gained from verbal evidence but this is not available here.

Level 2b is achieved, if you did not have batteries the light would not come on.

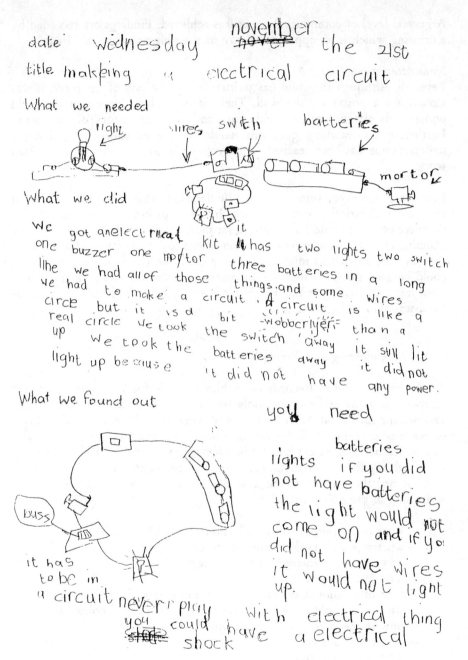

Figure 6.5: Example of pupil's work for an assessment folder

A general level of communication skill is achieved. Findings are recorded by a drawing which is an appropriate way of showing an electrical circuit.

New Attainment Target 5

Level 2b. Although the child has written this at the end of the piece, it was already on a poster on the wall. There is no real evidence that the child understands the nature of conductors and non-conductors of electricity. Further, more searching questions would be needed to establish that the understanding has been reached. This evidence is not available in the written work.

Level 3c is achieved, since he clearly understands that a complete circuit is needed for electrical devices to work. In terms of both NAT1 and NAT5, the piece of work could have allowed far more scope for the assessment. Our planning checklist in a previous section would assist the range of an activity such as this, so that as much as possible could be got from the activity. It could be used for some limited assessment as a clue for levels on the attainment target of writing in English, but it is not sufficiently detailed for a reliable assessment to be made.

3. Collecting and Storing Work

As the move towards Records of Achievement is fuelled by the National Curriculum, more and more schools are considering building up a portfolio of work for each child. Such portfolios of work should aim to provide direct evidence of progress, facilitate dialogue between teacher and child, child and parent, teacher and parent and by involving pupils directly in their learning and enhancing their self-esteem. However, before portfolios can be successfully built up, there needs to be a whole school policy. Issues which need addressing in the policy include:

- what is the purpose of a collection of work?
- what types of work should be stored?
- how are items entered into the collection?
- who puts them in?
- how often should the collection of work be reviewed?
- what is the role of pupils in choosing items for the collection?
- how is the collection kept current?
- where is it stored?
- who has access to it?
- when are items replaced or removed?

The agreement of all staff about questions such as these is important if portfolios of work are to be maintained. Ideally, a collection of work should serve two purposes. It should enable the teacher to maintain a record of evidence for herself and others about aspects of an individual's work and progress through the National Curriculum. It should also provide examples of the quality of work which an individual child wishes to retain. If both these purposes are adhered to, then several of the other questions have answers which are self evident.

The type of work stored should reflect the needs of the teacher for evidence regarding attainment in the National Curriculum, and the child's view of the quality work of which she or he is justifiably proud. Both the child and the teacher should have the right to enter work into a collection by negotiation. Only the teacher should have the right permanently to remove work from a collection. This should apply only if the work is to be replaced because the child has progressed to a higher level of attainment and thus the original has been superseded. Removal of work designated as quality by the child should not be removed by the teacher. Only the child has that right.

It will be important from the teacher's point of view that the collection is annotated. For example each piece of work could have written on the back:

— the context of the work recorded (done in a group, as an individual, as a result of a particular idea etc.),
— the reasons for selection (a good example of a particular level of attainment)
— the date of selection

Storage will always be a problem. We have referred earlier to methods of collection of evidence such as models and discussions. Some schools have developed hanging files in a portable trolley for the storage of work. Others have plastic box files in which a child's collection is stored. Whatever the method selected, the security of the evidence must be addressed as must the space collections take up in what are often already over-crowded classroom areas.

4. Role of Pupils in Assessment

Writing late in 1990, Dick West and Cathy Wilson defined some characteristics of a scheme of assessment that had pupils as the centre of the enterprise. (West and Wilson, 1990). It is very much in keeping with the earlier chapters of this book that pupils should be at the centre of the assessment process, and although there is no official place for them in the National Curriculum procedures as partners, rather as victims, we feel that the best practice should include them as equal partners in assessment.

West and Wilson say:

> First (a pupil-centred assessment scheme) would be planned as part of course design and ... pupils would be introduced to the assessment

criteria at the start of each unit or course ... Second, the pupils would be given the opportunity to test their progress at frequent intervals so as to use their self assessments in a formative manner. Third, adequate provision would be made for pupils to assess each other's work and to suggest suitable changes to assessment tasks and criteria.... Fourth, summative judgments would be reached through negotiation with peers and teachers and the results of all assessments would be incorporated in a Record of Achievement (profile). Finally, the whole process should place a high premium on recording positive achievements.

Some local education authorities (for example, Buckinghamshire, 1990) have prepared self-assessment sheets to help share the monitoring of progress with pupils. Their example below (Figure 6.6) has been prepared with reference to a unit of work, and draws from several of the old attainment targets in the 1989 science document within a given period of time. Pupils are provided with the sheets at the beginning of the unit, to be handed in at the end before a written test. At any point during the teaching the teacher can ask pupils to fill in some aspect of the sheet, or remind them to record progress. In practice teachers have used the sheets in other ways too. One, for example, made a point of targeting a small group of pupils every week during science. At some point during the lesson she would discuss their recorded progress on the sheet and raise with them questions about their level of confidence in understanding some of the more difficult issues. She also devoted one homework to the sheets towards the end of the unit, asking pupils to summarize their progress so far on their sheet. Part of the summary had to include evidence of how they knew they had achieved, understood etc. The second part of the summary included a list of issues they still found difficult and about which they felt insecure. From this homework, she would plan future lessons where groups could focus on particular aspects of difficulty, providing work for these groups to match their specific needs. Sometimes she would have to re-address an aspect of the work because the majority of the pupils had not understood. In this instance she designated 'experts' who did understand to work with different groups of pupils — acts of peer teaching. She found it took a good while to train the pupils to be honest about what they really understood, at both ends of the scale: some would not admit to any lack of understanding, while many were very prepared to say they understood nothing at all.

Recording Continuous Assessment

A variety of recording methods have been developed by different organizations and LEAs. We include a small collection here for those who do not wish to re-invent wheels. We have found the best to be those that are:

Communicating Achievement: Assessment in Science

Science Achievement Record — Knowledge and Understanding

Name _____ Class _____ Unit HEATING & FUELS

Old Attainment Target	During this unit I have learned the following	Pupil	Teacher	Assess
7.5.3	• I understand the process of combustion and respiration in which fuels and oxygen combine to form oxides.			
11.6.4	• I am able to read an electric meter and cost domestic electric energy consumption.			
13.4.1	• I understand that energy is essential to every aspect of human life and activity.			
13.4.2	• I know that there is a range of fuels which can be used to provide energy.			
13.5.1	• I understand the need for fuel economy and efficiency.			
13.5.2	• I understand the idea of global energy resources and appreciate that these resources are limited.			
13.7.1	• I understand energy transfer by conduction, convection and radiation in solids, liquids and gases. I understand the methods of controlling these transfers, particularly of insulation in domestic everyday contexts.			
13.7.3	• I am able to evaluate the methods used to reduce energy consumption in the home.			

Figure 6.6: Self-assessment sheet

— cheap
— easily reproduced
— simple and straightforward
— contain some simple quick reference to the statements of attainment
— provide space for comments against the levels of attainment.
— flexible so that the same information does not have to be written out in several different ways for several different audiences

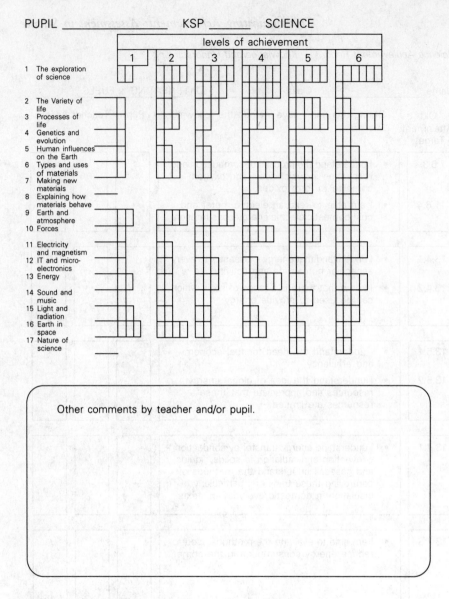

Figure 6.7: Recording continuous assessments: Science

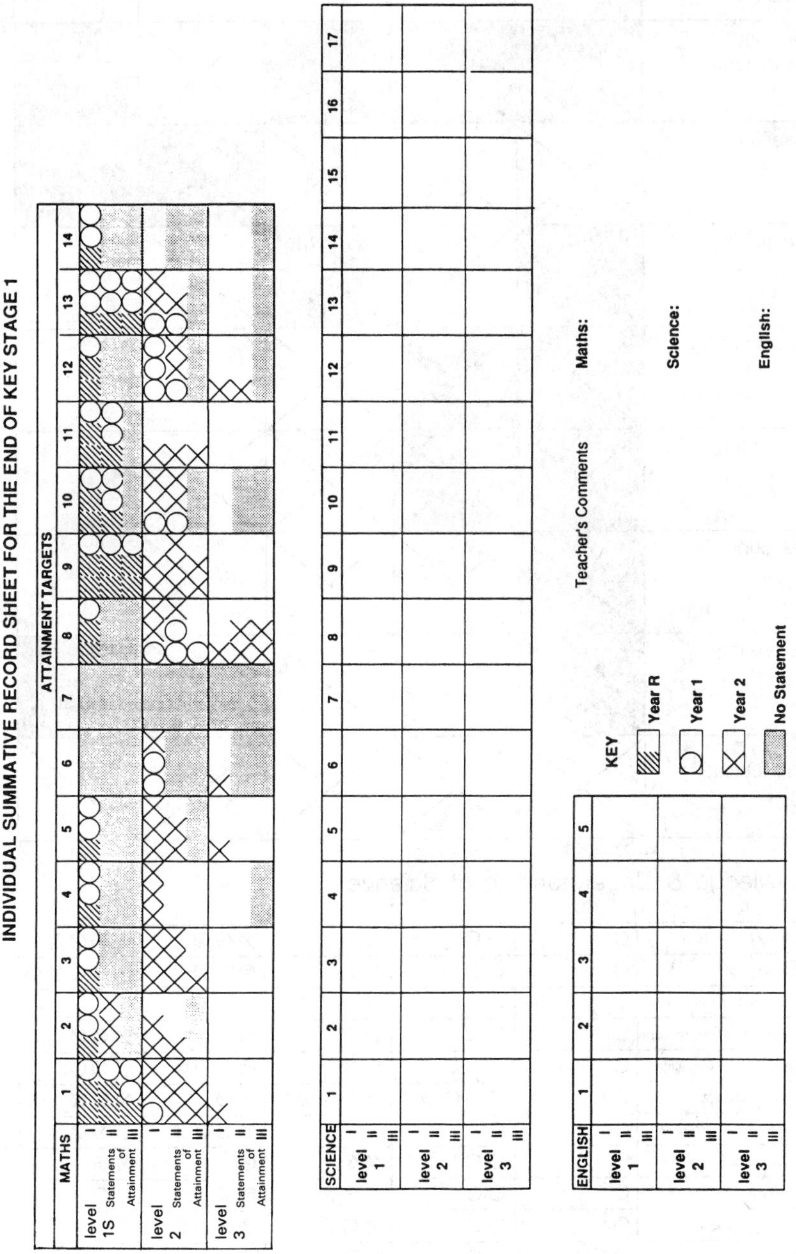

Communicating Achievement: Assessment in Science

Figure 6.8: Recording continuous assessment: Key Stage 1

Science National Curriculum Record

Name: /=Introduced & Covered   X = Does/knows with confidence
Date of Birth: *Please refer to Checklist Overleaf*

OAT1 Exploration of Science*	Level 1	Level 2	Level 3	Level 4	Level 5
Following Instructions					
Observing					
Questioning					
Hypothesising					
Fair Testing					
Measuring					
Interpreting					
Communicating					

Knowledge & Understanding of Science

	Level 1	Level 2	Level 3	Level 4	Level 5
AT2. The variety of life	O	O O	O O ⊠	⊠ O ⊠	⊠ ⊠ ∅ ⊠
AT3. Processes of Life	O	O O O	⊠ ⊠	⊠ ⊠ ⊠ O	⊠ ⊠ O ⊠ ∅
AT4. Genetics & Evolution	O	O	⊠	⊠	⊠
AT5. Human Influences on Earth	O	O O	⊠ ∅	⊠	⊠ ⊠ ∅
AT6. Types & uses of materials	O	O O O	O ⊠	⊠ ⊠ ⊠ ⊠ ∅	∅ O O
AT9. Earth & Atmosphere	O O	O O O O	∅ ⊠ ⊠ ⊠ ∅	⊠ ∅	⊠ ⊠ ⊠
AT10. Forces	O	O	⊠ ⊠	⊠ ⊠ ⊠ ⊠	O ⊠ ∅ ∅
AT11. Electricity & Magnetism	O	O O	⊠ ⊠	⊠	⊠ ⊠
AT12. Information Technology	O	O O	⊠ ⊠ ⊠	⊠ ⊠	O O
AT13. Energy Transfer	O O	O O	⊠ ⊠ ⊠	⊠ ⊠ O ⊠ ⊠	∅ ∅
AT14. Sound & Music	O	O O	⊠ ⊠	⊠	O O ∅
AT15. Light	O O	O O	⊠ ⊠	⊠ ⊠	O
AT16. The Earth in Space	O O O	O O O O	O O	⊠ ⊠ ⊠	O O

Figure 6.9: Assessment Records and Checklist for the 1989 Science National Curriculum

Old Attainment Target 1 Checklist

	Level 1	Level 2	Level 3	Level 4	Level 5
Following Instructions	Responds to direct verbal instructions in simple stages with frequent intervention.	Independently follows simple oral instructions.	Follows written instructions with occasional help.	Follows written instructions & diagrammatic representations. Carries out investigations with due regaard for safety.	Follows written instructions through a sequence of stages.
Observing	Observes using five senses.	Identifies simple differences: hot/cold, rough/smooth. Lists & collates observations	Selects and uses simple instruments to aid observations, eg. handlens, stopclock.	Chooses and competently uses a range of observation aids, including meassuring inssrtuments.	Identifies relevant observable features independently chooses equipment. Quantifies.
Questioning	Displays curiosity and asks questions	IAsks questions: How, why, what, if..? Suggests ideas	Able to ask a series of questions related to one subject or activity.	Raises questions in a form that can be investigated.	Uses concepts , knowledge and skills to suggest simple questions & designs investigations to answer them.
Hypothesising	Shows awareness of cause and effect. eg. 'if I push it it will roll'.	Tries to provide a simple explanation. eg. 'it falls because I drop it'.	Formulates hypotheses, eg. 'this ball will bounce higher than that one'.	Formulates testable hypothesis.	Provides some reasorting for hypothesis.
Fair Testing	Shows general awareness	Able to identify some variables with help	Distinguishes between fair & unfair test. Identifies & describes simple variables that change with time, eg. growth of a plant, burning candle, seasons.	Constructs fair tests. Plans investigations identifying & controlling variables.	Identifies & manipulates relevant independent & dependent variables, choosing appropriately between ranges, numbers & values.
Measuring	Aware of size and variation. Uses comparative words eg. bigger, louder	Uses non standard measures. Uses standard measures.	Quantifies variables, eg. measures the growth of a plant using a ruler. (Standard measure).	Selects & uses a range of measuring instruments to quantify physical quantities, eg. volume, temperature.	Selects & uses measuring instruments to quantify variables, and use more complex measuring instruments with more accuracy, eg. minor divisions on a thermometer.
Interpreting	Attempts to link cause and effect in events eg. 'in the morning the sun shines	Interprets findings, eg. 'light objects float' - 'thin wood is bendy'.	Interprets pictograms & bar charts. Interprets observation, eg. 'the greater the suspended weight, the longer the spring'.	Draws conclusions from experimental results.	Aware of limitations, eg. due to simple equipment identifies patterns in practical situations & number.
Communicating	Describes observations.	Records findings in charts, drawings etc.	Records in tables or bar charts. Describes activities carried out by sequencing major features.	Records results by appropriate mean, eg. simple tabes, bar charts, line graphs. Describes investigations in ordered prose, using limited technical vocabulary.	Makes written statements of the patterns derived from the data obtained from various sources.

Thanks to I.L.E.A.

Communicating in School Science

However, one aspect that it is important not to forget, is that recording of the National Curriculum levels is not the only recording task that needs to be done. Many teachers, particularly at the stage of transfer to another school, either in or at the end of a key stage, will want to record other aspects of achievement. SEAC have produced a useful guide for primary schools on Records of Achievement (SEAC, 1991) which has some pragmatic and helpful advice within it and is well worth reading. Several LEAs, in the absence of previous guidance from the DES have produced their own versions of records which can accompany a child on its progress from primary to secondary or between primary schools. Figure 6.10 below shows one such example — from Buckinghamshire.

Communicating (Reporting) Achievement

To Parents

Parents are entitled to know what level their child has achieved in any core subject in the National Curriculum. This information should be reported to them on a yearly basis, from continuous assessments, and at the end of the key stage, as a result of assessments plus SATs. They are also entitled to greater detail should they wish it — for example, the level in different profile components and different individual attainment targets. They will also receive information in the school brochure, about the school's performance generally in key stage assessments as their child progresses through that key stage.

To Parents in General and the Wider Community

Post Key Stage 1, schools are required to publish results of assessments to all parents. Full details of the publication required are not yet available, but clearly some indication of the following will be needed. Scores in science, maths and English for a year group, aggregated in some way, the mean, or the mode perhaps, plus the name of the school and some brief detail. It may be that, as with GCSE and A-level results in future, schools will be required to publish for parents details of the average scores obtained by children of that age in the rest of the LEA, and nationally. The DES will supply the latter information for the previous year for schools to use in their publication.

These are the basic minima. Clearly schools may wish to present more information. They may, for example, wish to present a 'change index' for parents. That is they may wish to inform parents what scores pupils had on entry at Key Stage 2, and what they had on exit, so that the effects of teaching are more obvious. This has advantages for those schools where, for example, bilingual pupils enter with little fluency in English, and by the school's

Communicating Achievement: Assessment in Science

PUPIL _____

KEY STAGE PROGRAMME   KSP [ ]
(unless specified)

Attendance during this school year:

Reasons for major absences:

CURRICULUM AREA	Achievement		in subject overall
	in profile component		
	Level *	Range *	AC
SCIENCE			
Exploration			
Knowledge /Understanding			
TECHNOLOGY			
MATHEMATICS			
Number, Algebra and Measures.			
Shape, Space, Data Handling.			
ENGLISH			
Speaking and Listening			
Reading			
Writing			
HISTORY			
GEOGRAPHY			
MODERN LANGUAGE(S)			
ART			
MUSIC			
RELIGIOUS EDUCATION			
PHYSICAL EDUCATION			

AC: Assessment Codes: V - Validated (final) } National Curriculum
                      E - Estimated (interim) } Levels
                      S - School's own curriculum programme

* : The numbered levels of achievement set out here represent summaries of more detailed information which the schools will let you have if given notice that you want it.

Figure 6.10

Communicating in School Science

efforts, achieve a great deal during Key Stage 2. Figure 6.11 below shows two pages taken from a school's brochures to parents as an illustration of both types.

Figure 6.11

THE WATTLEY ACADEMY
Parents will be interested in the results of the assessments for last years' Year 7 class, at the end of their Key Stage 2. As you know these children left us and went on to secondary schools in September.

The percentages shown are the percentages of the year 7 children who achieved the level shown. Where 0 is shown, there were no pupils operating at this level.

ENGLISH	reading	writing	spelling	handwriting
Level 2	0%	3%	5%	0%
Level 3	12%	24%	36%	10%
Level 4	47%	43%	52%	40%
Level 5	31%	22%	4%	37%
Level 6	10%	8%	3%	13%

SCIENCE	practical and scientific skills	scientific knowledge
Level 2	0%	0%
Level 3	17%	11%
Level 4	43%	37%
Level 5	31%	39%
Level 6	9%	13%

As parents can see, our children are achieving well in aspects of English and in science, where their factual knowledge is very good, and their practical skills excellent especially considering that primary schools are not equipped for science experimentation. To help parents make comparisons, we publish below the general achievement scores for schools in the borough in English and science at Key Stage 2.

ENGLISH	Borough schools	Our school
Level 1	0%	0%
Level 2	13%	2%
Level 3	28%	27%
Level 4	40%	37%
Level 5	14%	24%
Level 6	5%	10%

SCIENCE		
Level 1	0%	0%
Level 2	3%	0%
Level 3	33%	14%
Level 4	52%	40%
Level 5	10%	35%
Level 6	2%	11%

As parents can see the school is significantly better than the borough averages. More of our children are achieving at levels 5 and 6 than other schools in general.

MANFULLY PRIMARY SCHOOL
RESULTS OF YEAR 7 CHILDREN WHO LEFT US IN SEPTEMBER TO GO TO SECONDARY SCHOOLS
We think it is helpful to parents to see how this group of children were achieving when they came into the school at 7 years of age, in year 3 of their National Curriculum, so we have produced the results of their assessments at Key stage 1 (when they entered) and Key stage 2 (when they left us at eleven years of age).

ENGLISH	Key stage 1	Key stage 2
Level 1	22%	0%
Level 2	73%	5%
Level 3	5%	24%
Level 4	0%	57%
Level 5	0%	14%

SCIENCE		
Level 1	15%	0%
Level 2	71%	10%
Level 3	14%	43%
Level 4	0%	33%
Level 5	0%	9%
Level 6	0%	5%

As you can see, we have helped children to reach a great deal of their potential in the years they have been in Key Stage 2. To help parents see how successful the school is, we have given below the scores for children leaving our school at the end of Key Stage 2 alongside those from the borough schools in general at the end of Key Stage 2.

ENGLISH	Borough schools	Our school
Level 1	0%	0%
Level 2	13%	5%
Level 3	28%	24%
Level 4	40%	51%
Level 5	14%	14%
Level 6	5%	0%

SCIENCE		
Level 1	0%	0%
Level 2	3%	10%
Level 3	33%	43%
Level 4	52%	33%
Level 5	10%	9%
Level 6	2%	5%

As parents can see the school is significantly better than the borough averages in English. Considering that many of our children enter the school with little fluency, we feel that our hard work in the areas of language pay off in children's results. Our science results need looking at. We help some children to achieve very well — we have a higher percentage than the borough average at level 6, but as a spread, our children are not doing as well as they might at the lower levels. More of our children are still operating at level 3 than the borough average. Science is one of our development priorities for next year.

These are two very different ways of presenting the results. The first school is trading on the fact that for parents of their children, more detail about reading, writing and spelling achievements is important. They have tried to mitigate the fact that their results in profile component 2 are lower than those in profile component 1 by making statements about equipment. The Wattley Academy chooses to remind parents that the children improve dramatically in English during their time in the school, even though the science results on the whole are not as good as the rest of the borough. They do indicate that it is time to work on this however.

There is nothing in the regulations so far, to prevent schools presenting their results generally in these different ways, and thereby emphasizing their strengths and — perhaps — clouding their weaknesses. Manfully Primary School for example does not choose to give detailed results on reading, spelling etc., whereas the Wattley Academy deliberately tells parents about these achievements.

In reporting to other schools, a great deal of information will be required, and may need to be presented in a variety of different formats. Secondary schools, for example may need the information which goes to parents about an individual, i.e., at a level of generality, for the school file and tutor, whereas the heads of science, maths and English will require greater detail with regard to each attainment target. Formats such as those below may help here, as well as the transfer document detail in Figure 6.10.

Informing the Press

As the pressure for marketing schools becomes greater due to open enrolment and the Age-Weighted-Pupil-Unit formula method used for calculating LMS budgets for schools, publication of information to the press will no doubt be undertaken. What will be published remains to be seen. Schools may choose to publish individual pupil results at Key Stages 2 and 3 as many do for GCSE. They may wish to publish information about their school more generally. One can imagine the articles the Wattley Academy and the Manfully Primary School might write about their achievements.

WATTLEY ACADEMY — BEST SCHOOL IN THE BOROUGH!

The Wattley Academy's results in science and English at key stage 2 assessments far outshine those of the borough in general. More of our children achieve at higher levels than other schools in general. If parents want their children to have an education in which challenge and success are both present, this is clearly the school for them....

Communicating Achievement: Assessment in Science

BAKER JUNIOR SCHOOL

Form ...6f .....   School Year ..1989-90.

Levels of attainment

PUPIL	ENGLISH				MATHEMATICS			SCIENCE		
	Speak Listen	Read	Write	Subject	Number Algebra Measures	Shape Space DataHdlg	Subject	Explor- ation	Know Under- stand	Subject
Daniel Clash	5	5	4	5	5	5	5	4	4	4
Lucy Doban	6	5	4	5	4	4	4	4	4	4
Tessa Godley	5	5	4	5	4	3	4	3	3	3
Janice Moore	5	6	5	5	5	5	5	4	4	4
Roger Snow	6	6	6	6	6	6	6	5	5	5
Boyd Baker	5	4	3	4	4	4	4	5	5	5
Malcolm Bod	4	4	4	4	4	4	4	3	4	4
Lesley Book	3	4	3	3	3	3	3	3	3	3
Alan Buchan	5	6	5	5	4	4	4	4	4	4
Lew Clack	5	5	4	5	5	5	5	4	4	4
Tom Dainee	6	6	5	6	5	5	5	5	5	5
Ron Goss	4	5	4	4	4	4	4	4	4	4
Lanna Goy	3	3	2	3	2	2	2	2	2	2
Len Harrison	4	3	3	3	2	2	2	2	2	2
Colin Haswell	5	6	6	6	5	5	5	5	4	4
Helen James	5	5	5	5	4	4	4	4	4	4
Ron James	4	5	4	4	6	6	6	5	5	5
Vinny Key	6	4	3	4	3	4	3	4	4	4
Alan Ling	4	5	3	4	4	4	4	4	4	4
Frank Marsh	6	6	6	6	5	5	5	5	5	5
Steven Ord	2	3	2	2	2	2	2	2	2	2
Bill Oren	3	4	3	3	3	3	3	3	3	3
Agnes Orm	3	4	3	3	3	3	3	3	2	2
Janet Sinji	3	3	3	3	3	3	3	4	3	3
Stuart Steel	3	4	3	3	3	3	3	3	3	3
David Steave	3	3	2	3	2	2	2	2	2	2
Helen Walter	2	3	2	2	2	2	2	2	2	2
Shanee Wite	2	3	2	2	2	2	2	2	2	2
Weighting (%)	$33^{1}/_{3}$	$33^{1}/_{3}$	$33^{1}/_{3}$		60	40		45	55	

The weightings are those which are used to determine the subject level from the profile component levels.

ASE/Asst Seminar 90

Figure 6.12: Records of levels of attainment used by Buckinghamshire LEA

Communicating in School Science

> **PROGRESSING MANFULLY!**
>
> Children at Manfully Primary School progress more in four years than other schools in the borough. When pupils enter the school, their results are usually poor. Manfully helps them to progress to be just above average for the borough....

and so on.

These are clearly important areas of communication in which schools need to develop policies and wise practices. Successful communication in these important aspects of achievement represent potential parental interest and greater enrolment. Presenting the school in a favourable light is here to stay.

References

ABERCROMBIE, M.L.J. and TERRY, P.M. (1978) 'Talking to Learn: Improving teaching and learning in small groups', *Research into Higher Education Monographs*, 32, University of Surrey, Guildford.
ADAMS, J.L. (1974) *Conceptual Blockbusting. A guide to better ideas*, Penguin Books, Harmondsworth.
ADEY, P., BLISS, J., HEAD, J. and SHAYER, M. (1989) *Adolescent Development and School Science*, London, Falmer Press.
ALLPORT, F.H. (1924) *Social Psychology*, Boston, Houghton Miflin.
ARGYLE, M. (1969) *Social Interaction*, New York, Atherton Press.
ASPLEY, G. (1979) *Kids don't learn from teachers they don't like*, London, Academic Press.
ASSESSMENT OF PERFORMANCE UNIT (1984) 'Science in schools: Age 13', *Report*, 2, London, HMSO.
ASSOCIATION FOR SCIENCE EDUCATION (1979) *Alternatives for Science Education*, Hatfield, Association for Science Education.
ASSOCIATION FOR SCIENCE EDUCATION (1981) *A Statement of Policy*, Hatfield, Association for Science Education.
ASSOCIATION OF TEACHERS OF MATHEMATICS (1986)
ASSOCIATION FOR SCIENCE EDUCATION (1989) *Science and Technology in Society*, Hatfield, Association for Science Education.
AUSUBEL, D.P. (1964) 'How reversible are the cognitive and motivational effects of cultural deprivation? Implications for teaching the culturally deprived child', *Urban Education*, 1, pp. 32–41.
BALDWIN, J. and SMITH, A. (1983) *Active Tutorial Work Sixteen to Nineteen*, Blackwell, Oxford.
BARNES, D. (1976) *From Communication to Curriculum*, Harmondsworth, Penguin Books.
BARNES, D., BRITTON, J. and ROSEN, H. (1969) *Language, the Learner and the School*, Penguin, Harmondsworth.
BEATTIE, A. (1983) 'What Contribution Can Science Teachers Make to Health

References

Education?' Paper presented at the Annual Conference for the Association for Science Education, University of Exeter.

BELL, B. and FREYBERG, P. (1985) 'Language in the science classroom', in OSBORNE, R. and FREYBERG, P. (Eds) *Learning in Science*, London, Heinemann Educational.

BENTLEY, D. (1983) 'Teachers' hidden messages', *British Journal of Educational Psychology*, 53, pp. 121–27.

BENTLEY, D. (1985) '"It's not what you say ...": Youngsters' constructions of the meanings of teachers' non-verbal behaviour', Paper presented to the 6th International Congress on Personal Construct Psychology, Cambridge, August.

BENTLEY, D. (1987) 'Interviewing in the context of non-verbal research', in POWNEY, J. and WATTS, D.M. (Eds) *Interviewing in Educational Research*, London, Routledge, Kegan and Paul.

BENTLEY, D. and WATTS, D.M. (1989) *Learning and Teaching in School Science: Practical Alternatives*, Milton Keynes, Open University Press.

BRANDES, D. and GINNES, P. (1985) *A Guide to Student-Centred Learning*, Oxford, Basil Blackwell.

BRENT, LONDON BOROUGH (1989) *I Can Do That! Brent Primary Science*, Education Department, London Borough of Brent; London.

BUCKINGHAMSHIRE (1990) *Recording Achievement in Science*, Milton Keynes Curriculum Consortium Mimeograph, Buckinghamshire LEA.

THE BULLOCK REPORT (1975) *A Language for Life*, London, HMSO.

BUTTON, L. (1983) *Developmental Groupwork with Adolescents*, London, Hodder and Stoughton.

CARRE, C. (1981) *Language Teaching and Learning: Science*, London, Ward Lock Educational.

CASSELS, J.R.T. and JOHNSTONE, A.H. (1980) *Understanding Non-technical Words in Science*, Royal Society of Chemistry, London.

CASSELS, J.R.T. and JOHNSTONE, A.H. (1985) 'Words that matter in science', A report of a research exercise, The Royal Society of Chemistry, London.

CHARKIN, A., GILLEN, B., DERLEGA, V., HERNEH, J. and WILSON, M. (1983) 'Students' reactions to teachers' physical attractiveness and non-verbal behaviour: Two exploratory studies', *Psychology in the Schools*, 20, pp. 321–34.

CHILDREN'S LEARNING IN SCIENCE PROJECT (1990) *Teaching Schemes*, Centre for Science and Maths Education, Leeds, University of Leeds.

CLARK, C. (1985) 'Thoughts on the epistemological side effects of conceptual change teaching', Comments developed for the 1985 meeting of the invisible college of researchers on teaching, Chicago, March 1985.

CLARKE, K. (1991) Speech to the Annual Conference of the Association for Science Education, Birmingham University, January.

COGHILL, V. (1978) 'Infant school reasoning', Unpublished papers, Teachers' Research Group, University of London Institute of Education.

References

DEPARTMENT OF EDUCATION AND SCIENCE (1985) *Science 5–16: A Statement of Policy*, HMSO, London.
DEPARTMENT OF EDUCATION AND SCIENCE (1988) *The Report of the Working Party: Science*. London, DES.
DEPARTMENT OF EDUCATION AND SCIENCE (1989) *Science in the National Curriculum*, London, HMSO.
DEPARTMENT OF EDUCATION AND SCIENCE (1990) *Technology in the National Curriculum*, London, HMSO.
DITCHFIELD, C. (1987a) *Better Science 6: For Both Girls and Boys. Secondary Science Curriculum Review Curriculum Guides*, Heinemann Educational Books, London, and Association of Science Education, Hatfield, London, for the Schools Curriculum Development Committee.
DITCHFIELD, C. (1987b) *Better Science 7: Working for a Multicultural Society. Secondary Science Curriculum Review Curriculum Guides*, Heinemmann Educational Books, London, and Association of Science Education, Hatfield, London, for the Schools Curriculum Development Committee.
DRIVER, R. (1984) 'A review of research into children's thinking and learning in science', in BELL, B., WATTS, D.M. and ELLINGTON, K. (Eds) *Learning Doing and Understanding in Science: The Proceedings of a Conference*, London, Secondary Science Curriculum Review.
DRIVER, R. (1988) 'A constructivist approach to curriculum development', in FENSHAM, P. (Ed.) *Development and Dilemmas in Science Education*, London, Falmer Press.
DRIVER, R. and ERICKSON, G.L. (1983) 'Theories in action: Some theoretical and empirical issues in the study of students' conceptual frameworks in science', *Studies in Science Education*, 10, pp. 37–60.
DRIVER, R., GUESNE, E. and TIBERGHIEN, A. (1985) *Children's Ideas in Science*, Milton Keynes, Open University Press.
DRIVER, R. and OLDHAM, V. (1985) 'A constructivist approach to curriculum in science', Paper presented to the symposium 'Personal construction of meaning in educational settings', British Educational Research Association, Sheffield, August.
DRIVER, R., WATTS, D.M. et al. (1990) *Research on Students' Conceptions in Science: A Bibliography*, Children's Learning in Science Group, CSSME, Leeds, University of Leeds.
ENGINEERING COUNCIL (1985) *Problem Solving: Science and Technology in Primary Schools*, SCSST and Engineering Council.
ELLIS, R. (1984) *Classroom Second Language Development*, London, Pergamon Press.
FELDMAN, R.S. and ORCHOWSKY, S. (1982) 'Race and performance of students as determinants of teacher non-verbal behaviour', *Contemporary Educational Psychology*, 4, pp. 324–33.
FENSHAM, P. (1988) (Ed.) *Development and Dilemmas in Science Education*, London, Falmer Press.

References

FISHER, R. (1987) *Problem Solving in Primary Schools*, Basil Blackwell, Oxford.

FLANDERS, N.A. (1970) *Analysing Classroom Behaviour*, New York, Addison-Wesley.

FURNHAM, A.F. (1988) *Lay Theories. Everyday Understanding of Problems in the Social Sciences*, London, Pergamon Press.

FURTHER EDUCATION UNIT (1988) *Learning by Doing. A Guide to Teaching and Learning Methods*, London, DES.

GAGNE, R.M. (1970) *The Conditions of Learning*, New York, Holt, Rhinehart and Winston.

GALTON, M., SIMON, B. and CROLL, P. (1980) *Inside the Primary Classroom*, London, Routledge and Kegan Paul.

GASKELL, G. and SEALY, P. (1976) *Groups. Social Psychology Course D305*, Milton Keynes, Open University Press.

GAULD, C. (1989) 'A study of pupils' responses to empirical evidence', in MILLAR, R. (Ed.) (1989) *Doing Science: Images of Science in Science Education*, London, Falmer Press.

GILBERT, G.N. and MULKAY, M. (1984) *Opening Pandora's Box. A Sociological Analysis of Scientists' Discourse*, Cambridge University Press.

GILBERT, J.K. and WATTS, D.M. (1983) 'Concepts, misconceptions and alternative conceptions: changing perspectives in science education', *Studies in Science Education*, 10, pp. 61–91.

GILBERT, N.S. (1987) 'Solving problems in science', *Education 3–13*, pp. 21–4.

GILL, D. and LEVIDOW, L. (1987) (Eds) *Anti-racist Science Teaching*, London, Free Association Books.

GRADED ASSESSMENT IN SCIENCE PROJECT (1987) London, ILEA.

GREENE, J. (1975) *Thinking and Language*, London, Methuen.

HADFIELD, J.M. (1987) 'Problem-oriented structured teaching', *Education in Chemistry*, March, pp. 43–4.

HART, S. (1989) 'Collaborative Learning, Groupwork and the Spaces in Between', Mimeograph, Thames Polytechnic, London.

HOLT, G.R. (1990) '"Doing science in different voices": Bakhtin's Heteroglossia and the analysis of the discourse of multi-disciplinary researchers', Paper presented to the International Communication Association Annual Meeting, Dublin, June 1990. Department of Speech Communication, University of Illinois.

HORSCROFT, D. and POPE, M.L. (1985) *Students and Teachers*, Module A2 Study Guide, Diploma in the Practice of Science Education, University of Surrey/Roehampton Institute.

HOUSTON, G. (1984) *The Red Book of Groups*, Houston, Norfolk.

HOWARD, R.W. (1987) *Concepts and Schematas: an Introduction*, Cassell Educational, London.

JACKSON, K.F. (1983) *The Art of Problem Solving: Bulmershe-Comino Problem Solving Project*, Reading, Bulmershe College.

References

JOHNSON, D.W. and JOHNSON, R.T. (1985) *Learning Together and Alone*, Englewood Cliffs, NJ: Prentice-Hall.

KAHNEY, H. (1986) *Problem Solving: A Cognitive Approach*, Milton Keynes, Open University Press.

KELLY, A. (1987) *Science for Girls?* Milton Keynes, Open University Press.

KELLY, G.A. (1955) *The Psychology of Personal Constructs. Volumes 1 and 2.* New York, W.W. Norton and Co.

KINGDON, J.M. and CRITCHLEY, W.E. (1982) 'The use of technical terms in CSE and O-Level chemistry', *School Science Review*, 64, 227, pp. 367–72.

LIPPITT, J. and WHITE, R. (1960) *Autocracy and Democracy*, New York, Harper and Row.

LYTHE, M. (1986) *How Scientists Work. Nuffield 11–13 Science*, London, Longman, for Nuffield Chelsea Curriculum Trust.

MCQUAIL, D. and WINDAHL, S. (1981) *Communication Models for the Study of Mass Communication*, London, Longmans.

MAHONEY, M.J. (1988) 'Constructive meta-theory', *International Journal of Personal Construct Psychology*, 1, 1, pp. 1–36.

MAIER, N.R.F. (1963) *Problem Solving Discussions and Conference: Leadership Methods and Skills*, New York, McGraw-Hill.

MARLAND, M. (1975) *The Craft of the Classroom. A Guide to Survival*, London, Heinemann Educational.

MARSHALL, S., GILMOUR, M. and LEWIS, D. (1990) 'Understanding and misunderstanding in science', Research Report No. 18, Department of Language and Communication Studies, University of Technology, Lae, Papua New Guinea.

MARTIN, N. et al. (1976) *Writing and Learning Across the Curriculum*, Harmondsworth, Penguin Books.

MERMIN, N.D. (1990) *Boojums All the Way Through: Communicating Science in a Prosaic Age*, Cambridge University Press.

MINSTRELL, J. (1982) 'Explaining the 'at-rest' condition of an object', *Physics Teacher*, 20 January, pp. 10–14.

NATIONAL CURRICULUM COUNCIL (1989) *Science Non-statutory Guidance*, York, National Curriculum Council.

NATIONAL CURRICULUM COUNCIL (1990) *Curriculum Guidance 3: The Whole Curriculum*, London, Dept. of Education and Science, HMSO.

NATIONAL CURRICULUM COUNCIL (1991) *Science for Ages 5–16*, London, Dept. of Education and Science, HMSO.

NIXON, J. and WATTS, D.M. (1989) *Whole School Approaches to Equal Opportunities, Inset Workshops for Schools*, Basingstoke, Macmillan.

OGBORN, J. (1991) 'Science and English', in WATTS, D.M. (1991) (Ed.) *Science in the National Curriculum*, London (in press), Cassells.

OSBORNE, R. and FREYBERG, P. (1985) *Learning in Science*, London, Heinemann Educational.

PHYSICS EDUCATION (1985) Volume 20.

References

PINES, M. (1975) 'Overview', in KREEGER, L. (Ed.) *The Large Group*, London, Constable.

PLAYFOOTS (1990) 'Problem solving in primary science', Unpublished MA dissertation, London, Roehampton Institute.

POPE, M.L. and WATTS, D.M. (1988) Constructivist 'Goggles: Implications for process in teaching and learning physics', *European Journal of Physics*, 9, pp. 101–109.

RICE, W. (1981) *Informal Methods in Health Education*, Teachers' Advisory Council for Drugs and Alcohol Education (TACADE), Manchester.

RICHARDS, J. (1979) *Classroom Language: What Sort?* London, George Allen and Unwin.

RICHEY, A. and RICHEY, B. (1982) 'Non-verbal behaviour in the classroom', *Psychology in the Schools*, 19, pp. 224–231.

ROACH, T., SMITH, D. and VAZQUEZ, M. (1990) *Language and Learning in the National Curriculum: Bilingual Pupils and Secondary Science*. Hounslow, Schools Language Centre.

ROSENSHINE, L. (1970) 'Enthusiastic teaching: a research review', *School Review*, August, pp. 499–514.

RYDER, R.D. (1975) *Victims of Science. The Use of Animals in Research*. London, Anti-vivisection League.

SCSST (1985) *Problem Solving: Science and Technology in Primary Schools*, SCSST and Engineering Council.

SEAC (1989) *A Guide to Teacher Assessment, Pack B. Teacher Assessment in the School*, London Schools Examinations and Assessment Council.

SEAC (1990) *National Curriculum Assessment: LEA Responses: A report*. London, School Examinations and Assessment Council.

SEAC (1991) *Records of Achievement: A Guide for Primary Schools*, London, School Examinations and Assessment Council.

SCIENCE IN PROCESS (1987) Heinemann Educational in association with the ILEA, London.

SECONDARY SCIENCE CURRICULUM REVIEW (1984) *Towards the Specification of a Minimum Entitlement: Brenda and Friends*, London, Secondary Science Curriculum Review.

SEMIN, G.R. and GERGEN, K.J. (1990) *Everyday Understanding: Social and Scientific Implications*, London, Sage Publications.

SHAYER, M. and ADEY, P. (1981) *Towards a Science of Science Teaching*, London, Heinemann Educational.

SOLOMON, J. (1991) 'School laboratory life', in WOOLNOUGH, B. (1991) (Ed.) *Practical Science: The Role and Reality of Practical Work in School Science*, Milton Keynes, Open University Press.

STEWART, D. (1987) *Better Science 7: Making it Relevant*, Secondary Science Curriculum Review Curriculum Guides, Heinemann Educational Books, London, and Association of Science Education, Hatfield, for the Schools Curriculum Development Committee, London.

STRIKE, K.A. and POSNER, G.J. (1985) 'A conceptual change view of learning and understanding', in WEST, L.H.T. and PINES, A.L. (Eds) *Cognitive Structure and Conceptual Change*, London, Academic Press.

SWIFT, D.J. (1984) 'Against structuralism: is genetic epistemology a conservative activist theory of knowledge?' Paper presented to 10th Annual Conference of British Educational Research Association, Lancaster, August.

SUTTON, C. (1981) (Ed.) *Communicating in the Classroom*, London, Hodder and Stoughton.

TRIPLETT, N. (1897) 'The dynamogenic factors in pacemaking and competition', *American Journal of Psychology*, 9, pp. 503–533.

TURQUET, L. (1975) 'Threat to Identity in Large Groups', in KREEGER, L. (Ed.) *The Large Group*, London, Constable.

VYGOTSKY, L. (1986) *Thought and Language*, (Revised edition) Cambridge, Massachusetts, MIT Press.

WALLACE, J. (1986) Social 'Interaction within Second Year Groups', Unpublished MSc Thesis, University of Oxford.

WATTS, D.M. (1983) 'A study of alternative frameworks in school science', Unpublished PhD Thesis, University of Surrey.

WATTS, D.M. (1989) *The Egg-race Factfile*, Surrey Satro and Esso, University of Surrey.

WATTS, D.M. (1989) 'Discussing physics', in BENTLEY, D. and WATTS, D.M. (1989) *Teaching and Learning in School Science: Practical Alternatives*, Milton Keynes, Open University Press.

WATTS, D.M. (1991) *The Science of Problem Solving*, London, Cassell Educational.

WATTS, D.M. and BENTLEY, D. (1986) 'Constructivism in the classroom: enabling conceptual change by words and deeds', *British Educational Research Journal*, 13, 2, pp. 121–135.

WATTS, D.M. and GILBERT, J.K. (1983) 'Concepts, misconceptions and alternative frameworks: changing perspectives in school science', *Studies in Science Education*, 10, pp. 61–91.

WATTS, D.M. and GILBERT, J.K. (1986) 'Appraising students' understanding of concepts in physics. Gravity', Mimeograph, British Petroleum and University of Surrey Department of Educational Studies.

WATTS, D.M. and GILBERT, J.K. (1989) 'The new learning: Research, development and the reform of school science education', *Studies in Science Education*, 16, pp. 75–121.

WATTS, D.M. and O'BRIEN, E.M. (1989) 'Developing an equal opportunities policy on gender', *Inset Workshops for Schools*, Basingstoke, Macmillan.

WATTS, D.M and POPE, M.L. (1985) 'Modulation and fragmentation: some cases from science education', Paper presented to the 6th International Congress on Personal Construct Psychology, Cambridge, August.

WATTS, D.M. and POPE, M.L. (1989) 'Thinking about thinking, learning about learning: Constructivism in physics education', *Physics Education*, 24, pp. 326–331.

References

WELLINGTON, J. (1983) 'A taxonomy of scientific words', *School Science Review*, 64, 229, pp. 767–773.

WEST, L.H.T. and PINES, A.L. (1985) *Cognitive Structure and Conceptual Change*, London, Academic Press.

WEST, R.W. and WILSON, C. (1990) *Assessment and Accountability*. (ASE Assessment Seminar) Student based or student related assessment. Mimeograph Hatfield, Association for Science Education.

WHITE, R. (1989) *Learning in Science*, Basil Blackwell, Oxford.

WHITE, S. (1990) 'What children understand about heat and how this relates to the National Curriculum', in BARBER, B. and WATTS, D.M. (Eds) *Doing the Difficult Bits*, An occasional publication, Roehampton Institute (in press).

WORLD WIDE FUND FOR NATURE (1989) *The Environmental Enterprise Award Scheme*, WWF (UK), Godalming.

WILKINSON, A. (1971) *The Foundation of Language*, Oxford, Oxford University Press.

WILKINSON, A., STRATTA, L. and DUDLEY, P. (1974) *The Quality of Listening*, Schools Council Research Studies. Macmillan, Basingstoke.

YEOMANS, R. (1988) 'Making the large group feel small: Primary teachers' classroom skills, *Cambridge Journal of Education*, 17, 3.

Author and Subject Index

Abercrombie 45
Adams 126
Adey 16, 34
Allport 62
alternative frameworks 15–16, 141
AOT (adult other than teacher) 1, 44, 126, 136
APU (assessment of Performance Unit) 112
Argyle 29
ASE (Association for Science Education) 5, 7, 69, 140
Aspley 37
assessment 138–182
   continuous 143, 152, 155–156
   evidence for 155–168
   pupil centred 170
   self 26, 145, 153, 171
   whole class 72, 74
ATM (Association of Teachers of Mathematics) 118
auditing 114–115
Ausubel 37

BA (British Association) 116
Baldwin 60
Barnes 5, 42
Beattie 45
Bell 10
Bentley 18, 29, 38, 113, 118
bilingualism 49–55, 70, 176

Bliss 183
brainstorming 87, 135
Brandes 66, 110, 111
Brent (London Borough of) 116
Britton 42
Buchinghamshire (local education authority) 170, 176
Bullock Report 5
Button 56

Carre 5
Cassels 10
Charkin 38
Clark 36
CLIS (Childrens Learning in Science Project) 16, 34
CLSS (Community Language Support Service — Hounslow) 16, 34
Coghill 15
computers 19, 20
concept development 4, 11, 12, 16, 31, 34–36, 41, 56, 60, 63, 65, 78
constructivism 4, 11–17, 33–36, 65, 102
CPVE (Certificate of Pre Vocational Education) 112
creativity 75, 121
CREST (Creativity in Science and Technology) 116–117

191

Critchley 10
Croll 187

DES (Department of Education and Science) 3, 22, 23, 112, 113, 140, 176
diary 93–94
Derlega 184
discussion, 17, 36, 42–43, 51, 69, 94, 105, 118, 124, 133, 145, 156, 164
display 30–31, 36, 69, 97, 135
Ditchfield 65, 66
Driver 10, 16, 33, 34, 35, 36, 37, 106

Earth 12, 69, 105–106
Ellington 185
Ellis 49
Emery 90
Engineering Council 113
English National Curriculum 3, 20, 44, 94–95, 140, 148, 178, 179
environment 4, 27, 30, 118, 142
  local 97, 126, 145, 159
  non-threatening 27–30, 33, 34, 66, 87, 88, 135
  physical 30, 32–33, 44, 111
equal opportunities
  gender 31, 32, 43, 62, 64–72, 96, 97, 103, 106
  race 31, 32, 62, 64–74, 97, 106
Erickson 34
expectation 18, 102
exploration 18, 101, 102, 133
explication 18, 101, 102, 133

Feldman 29
Fensham 16
FEU (Further Education Unit) 47
Fisher 114
Flanders 42
Freyberg 10, 16, 36
Furnham 16

GASP (Graded Assessment in Science Project) 76
Gagne 117
Galton 59
Gauld 19
Gaskell 62
GCSE (General Certificate of Secondary Education) 3, 58, 77, 139, 143, 154, 155, 156, 176, 180
Gergen 16
Gilbert GN. 9
Gilbert JK. 3, 10, 16, 23, 34, 80
Gilbert NS. 118
Gill 66
Gillen 184
Gilmour 10
Ginnis 65, 110, 111
graphics 19, 135–136, 145
groups
  definitions 57, 58
  dynamics 62, 87–88, 90–91, 110, 124
  large 60, 61, 67–68
  leaders 83–87
  norms 63, 87–88, 122
  pairs 61, 62, 134
  peer 57, 65, 144–145
  random 67–68, 70
  small, 56, 64, 88–89
Guesne 185

Hadfield 117
Hart 28, 87, 88
Head 183
Hermeh 184
Holt 9
Horscroft 33
Hounslow Language Support services 49, 50
Houston 84
Howard 11
hypothesising 21, 68, 74–75, 150–151, 174, 175

ILEA 76
images 31, 32

Jackson 113
Johnson 81
Johnstone 10

Kahney 113
Kelly A 65
Kelly GA 12
Key Stages
  KS1: 1, 6, 19, 70, 74, 77, 93–96, 139, 140, 146, 148, 156, 157, 166, 173, 176
  KS2: 19, 45, 46, 67, 70, 74, 77, 78, 81, 97, 98, 100, 114, 126, 133, 134, 146, 148, 156, 176, 177, 180
  KS3: 20, 44, 46, 75, 77, 78, 81, 102, 105, 114, 122, 128, 141, 145, 146, 154, 155, 156, 159, 180
  KS4: 20, 32, 42, 75, 77, 107, 109, 114, 130, 145, 146, 154, 156
Kingdon 10
Kreeger 187

Lewis 10
Levidow 66
Lippitt 83
listening 45–49, 50, 134

Mahoney 11
Maier 84
Marland 28
Martin 5
Mathematics National Curriculum 3, 20, 72, 148, 157
Marshall 10
McQuail 7
Mermin 8
Miller 187
mime 99–100, 100, 102, 156
Minstrell 33
models 19, 135, 136, 156

Mulkay 9
music 17, 100–101

NAT (New Attainment Targets)
  NAT1: 20, 67, 72, 118–120, 143, 147, 148, 149, 150, 151, 157, 158, 159, 160, 163, 164, 166, 168
  NAT2: 93, 95, 97, 98, 159
  NAT3: 69, 94, 105, 109, 155, 158
  NAT4: 81, 102, 146, 148
  NAT5: 100, 107, 150, 168
NCC (National Curriculum Council) 3, 5, 6, 17, 19, 22, 42, 49, 56, 77, 92–111, 112, 117, 119–120, 136, 142
Nixon 66
NFER (National Foundation for Educational Research) 138
non-verbal
  behaviour 29, 36–42, 47
  communication 18, 27, 28
numeracy 16

O'Brien 65
Ogborn 42
Oldham 33, 36, 37
ORACLE 59, 60
Orchowsky 29
Osborne 16, 36

Physics Education 34
Piaget 16
Pines 34, 61
Playfoot 44
plays 99–100
poetry 94, 97, 136, 156
Pope 12, 33, 36
Posner 35, 36, 41
problem solving 4, 44, 107–108, 112–137
progression 23, 25–26, 153
project work 64, 125

recording 93–94
reporting 97, 133–134
Rice 88
Richards 10
Richey 29
Roach 49, 50
roles 80–87, 107–109, 123, 127, 128, 130–131
role-play 122
Rosen 42
Rosenshine 37
Ryder 43

SAT (Standard Attainment Task) 57, 138, 140, 141
SATIS (Science and Technology in Society) 69
SATRO (Science and Technology Regional Organisation) 116, 117, 122, 124
Science in Process 76, 87
SCSST (Standing Committee for School Science and Technology) 113
SEAC (Schools Examination and Assessment Council) 139, 141, 142, 152, 176
Sealy 62
Semin 16
Shaw 112
Shayer 34
Simon 187
skills 25, 63, 64, 67–68, 74–78, 92–111, 120, 138, 139
 teacher skill 41–42
 skill teaching 42–49, 72–78, 125, 133–137
 skill wheel 73, 74, 77–78
Smith A 60
Smith D 188
Solomon 44
SSCR (Secondary Science Curriculum Review) 7, 34, 112
Stewart 112, 113

Strike 35, 36, 41
Swift 35
Sutton, 5, 46

TACADE 88
Technology National Curriculum 157
Terry 45
themes 3
Tiberghien 185
2D (three dimensional) 31, 121, 157, 158, 164
Triplett 61
3D (two dimensional) 31, 113, 121, 158, 164
Turquet 61

Vygotsky 11, 12, 56
Vazquez 188

Walker 109
Wallace 44
Watts 3, 10, 12, 16, 23, 34, 36, 44, 65, 66, 80, 113, 118, 128
West L. 34
West R. 169, 170
Wellington
Wilkinson A 45
Wilkinson S 42
Wilson C 169, 170
Wilson M 184
Windahl 10
White R (1988) 16, 34
White R (1960) 83
White S 166
Woolnough 188
worksheets 45–46, 82, 104, 168–173
writing 51, 93, 136, 145, 166
WWF (World Wide Fund for Nature) 117

Yeomans 29, 60, 61, 88, 90
Young Investigators 116

The South East Essex
College of Arts & Technology
Carnarvon Road, Southend-on-Sea, Essex SS2 6LS
Phone 0702 220400 Fax 0702 432320 Minicom 0702 220642